D1545395

JN 8-11-80

INFORMATION SOURCES
FOR RESEARCH AND DEVELOPMENT

Use of
Chemical Literature

INFORMATION SOURCES
FOR RESEARCH AND DEVELOPMENT

A series under the General Editorship of

R.T. Bottle, B.Sc., Ph.D., C.Chem., F.R.I.C., F.L.A., F.I.Inf.Sci.,
and
D.J. Foskett, M.A., F.L.A.

Other titles in the series

Use of Biological Literature (2nd edition)
    edited by R.T. Bottle & H.V. Wyatt

Use of Criminology Literature
    edited by M. Wright

Use of Earth Sciences Literature
    edited by D.N. Wood

Use of Economics Literature
    edited by J. Fletcher

Use of Engineering Literature
    edited by K.W. Mildren

Use of Management and Business Literature
    edited by K.D.C. Vernon

Use of Mathematical Literature
    edited by A.R. Dorling

Use of Medical Literature (2nd edition)
    edited by L.T. Morton

Use of Physics Literature
    edited by H. Coblans

Use of Reports Literature
    edited by C.P. Auger

Use of Social Sciences Literature
    edited by N. Roberts

Use of
Chemical Literature

Third Edition

Editor

R. T. Bottle, B.Sc., Ph.D., C.Chem., F.R.I.C., F.L.A., F.I.Inf.Sc.
Professor of Information Science,
Centre for Information Science,
The City University, London

BUTTERWORTHS
LONDON - BOSTON
Sydney - Wellington - Durban - Toronto

The Butterworth Group

| United Kingdom | Butterworth & Co (Publishers) Ltd |
| | London: 88 Kingsway, WC2B 6AB |

Australia   **Butterworths Pty Ltd**
Sydney: 586 Pacific Highway, Chatswood, NSW 2067
Also at Melbourne, Brisbane, Adelaide and Perth

Canada   **Butterworth & Co (Canada) Ltd**
Toronto: 2265 Midland Avenue, Scarborough,
Ontario M1P 4S1

New Zealand   **Butterworths of New Zealand Ltd**
Wellington: T & W Young Building, 77–85 Customhouse
Quay, 1, CPO Box 472

South Africa   **Butterworth & Co (South Africa) (Pty) Ltd**
Durban: 152/154 Gale Street

USA   **Butterworth (Publishers) Inc.**
Boston: 10 Tower Office Park, Woburn, Mass. 01801

First published	1962
Second edition	1969
Reprinted with revisions	1971
Third edition	1979

© Butterworth & Co (Publishers) Ltd, 1979

ISBN 0 408 38452 2

British Library Cataloguing in Publication Data

Use of chemical literature. – 3rd ed. –
   (Information sources for research and development).
   1. Chemistry – Information services
   2. Chemistry – Bibliography
   I. Bottle, Robert Thomas II. Series
   540'.7   QD8.3   79-41061

   ISBN 0–408–38452–2

Typeset by Scribe Design, Medway, Kent
Printed in England by Whitefriars Press, Tonbridge, Kent

Preface to the third edition

Many changes in the chemical literature have taken place in the decade since the second edition was written. The electronic journal is virtually with us as a means of disseminating primary information and it certainly exists as an alternative to many abstracts journals. Indeed it is the only format of one or two services which previously existed in print. Sections on computerized services and networks have therefore been added to the Abstracts chapter. A brief introduction to chemical coding and Wiswesser Line-Formula Notation has been included with material on nomenclature and compound indexes in Chapter 6. This is not intended to turn readers into expert WLN encoders, but merely to stimulate some to seek further instruction from the sources discussed there.

Even the Beilstein chapter (11) has some extra material added, though it is still the one which is least changed from the first edition. Major changes in the Patents scene (and a new contributor) have caused this chapter to bear little resemblance to that in previous editions. Pressure on space did not permit the inclusion of an acronyms glossary in this edition, as it was estimated that this would have at least doubled in size. Some of the more important ones, however, may be decoded via the Index.

I welcome several new contributors to this edition: Mrs Kostrewski and Mrs Rennie, Drs Belkin, Jones and Oppenheim and Messrs Osborne and Sweeney. I should like to thank contributors to previous editions for their past help and am sorry to record the recent death of one of them, Mr J. Lucas, who wrote the nuclear chemistry chapter for the first and second editions.

I again thank the numerous colleagues and students who made helpful comments on the second edition, many of which I have been able to incorporate in this edition.

London   R.T. Bottle

Extracts from the Preface to the second edition

This book marks the formal start of a new series of literature guides entitled *Information Sources for Research and Development* which will be under the general editorship of D.J. Foskett and myself. The many changes in the nature and pattern of the chemical literature since 1962 have necessitated an almost complete rewriting of most of the chapters — indeed only the chapters on Beilstein and on Patents are substantially the same. Two new chapters, Chapter 12* (Polymer Science) and Chapter 15* (Some Less Conventional Sources of Information) have been added, though the last part of Chapter 15* is based on the Research Intelligence section which appeared in Chapter 6 in the first edition. New sections on spectroscopic data and on general physical chemical works have been added to Chapter 7*, whilst a 'physical methods' section has been included in Chapter 11.* One on interdisciplinary indexes has been added to Chapter 4* and one on chemical prices and suppliers to Chapter 14.* A section on analytical chemistry has been appended to Chapter 8,* whilst specifically organic aspects are dealt with in Chapter 11.* Several new exercises have been added to widen the scope of the practical work which is so essential in any course on chemical literature. It is hoped that these and other changes will increase the scope and usefulness of the book.

Inevitably pressure on space has caused certain works to be omitted. The editor and contributors have tried to include those sources which they find most useful rather than presenting an uncritical mass of literature from which the reader is left to make his own selection. Ultimate responsibility for errors of omission or commission, however, rests with the editor rather than with the contributors. To save needless

*The numbers of the corresponding chapters in the 3rd edn are increased by 1.

repetition and space, the town of publication has generally been omitted in the case of major British, American and international publishers; such information may readily be found in the standard reference works (see Chapter 2).

Finally thanks are due to numerous colleagues, both chemists and information workers, for helpful suggestions, information, etc.

Syracuse, N.Y.   R.T. Bottle

Extracts from the Preface to the first edition

Many books have their origin in a course of lectures. Most of the ensuing chapters are based on lectures given at the short courses organized by Liverpool College of Technology in collaboration with LAD-SIRLAC during the past three years. Although tabular matter and many references which are difficult to put in lectures have been included, book lists and similar material have been kept to the minimum in order to make the book as readable a text as possible. In many cases examples only have been given since the aim of comprehensiveness is often self-defeating and 'comprehensive lists' soon get out of date. Every effort has been made in selecting the material and in the layout of the book to make it a useful reference tool.

Practice is the key to the efficient use of the chemical literature and thus a selection of practical exercises with notes on their solution are appended. These have been tested out at the Liverpool courses and enable the new graduate to get the practice he needs in using the literature before tackling his own important literature problems.

As this book is the product of many authors, the reader will notice differences in style and presentation in the various chapters. Furthermore he will observe that full and formal bibliographical citations have not been given for books etc. referred to in the text. This has been done, of course, to conserve space and improve readability.

My first acknowledgement must be to the late Dr A. Kronenberger, formerly Chemistry Librarian at Indiana University, whose courses first convinced me of the need for formal instruction in the use of the chemical literature. I am also indebted to Mr B.C. Vickery for helpful discussions. My task as editor has been made easier by the generous help of librarian colleagues, in particular Messrs C.R. Burman, C.A. Crossley and F. Earnshaw. The late Mr K. Hopkins very kindly gave me his lecture notes on Government and trade publications on which I have based Chapter 13 [Chapter 15 in the 3rd edn]. Dr R.E. Fairbairn also kindly made available to me his lecture notes on abstracts, abstracting and information retrieval which have formed the basis of Chapter 5. Many of the contributors devised exercises relevant to their chapters, but for Exercise 2 and part of Exercise 3, I am indebted to Mr T.N. Shaw of Unilever Ltd. The Chemical Abstracts Service has been most helpful in supplying information, answering queries and permitting reproduction of journal abbreviations. Finally I would like to express my gratitude to my wife, Margaret, for help with the manuscript and the proofs and for her constant encouragement.

Bradford   R.T. Bottle

Contributors

N.J. BELKIN, B.A., M.L., Ph.D., Centre for Information Science, The City University, London

R.T. BOTTLE, B.Sc., Ph.D., C.Chem., F.R.I.C., F.L.A., F.I.Inf.Sc., Professor of Information Science, Centre for Information Science, The City University, London

C.R. BURMAN, B.A., F.L.A., Librarian, University of Aston, Birmingham

F. EARNSHAW, B.A., A.L.A., Librarian, University of Bradford

R.B. HESLOP, M.Sc., C.Chem., F.R.I.C., Chemistry Department, University of Manchester Institute of Science and Technology

K. JONES, Ph.D., C.Chem., M.R.I.C., Chemistry Department, University of Manchester Institute of Science and Technology

B.J. KOSTREWSKI, B.Sc., M.Sc., M.I.Inf.Sc., Centre for Information Science, The City University, London

C. OPPENHEIM, B.Sc., Ph.D., M.I.Inf.Sc., Centre for Information Science, The City University, London

A.G. OSBORNE, M.Phil., C.Chem., M.R.I.C., Chemistry Department, The City University, London

T.C. OWEN, B.Sc., Ph.D., C.Chem., F.R.I.C., Chemistry Department, University of South Florida, Tampa

J.S. RENNIE, B.Sc., F.R.C.S., M.A., M.I.Inf.Sci., Centre for Information Science, The City University, London

R.M.W. RICKETT, B.Sc., Ph.D., C.Chem., F.R.I.C., Director, Middlesex Polytechnic, London

J.M. SWEENEY, B.Sc., M.Sc., A.I.Inf.Sc., British Technology Index, London

J.L. THORNTON, F.L.A., Consultant Librarian, Royal College of Obstetricians and Gynaecologists, London

R.I.J. TULLY, F.L.A., Deputy Librarian, University College of North Wales, Bangor

B. YATES, B.Sc., F.I.Inf.Sc., National Library of Australia, Canberra

Contents

1

Introduction

R. T. Bottle

Chemists can communicate directly by correspondence, by visits and at conferences. On most occasions, however, they communicate indirectly through the literature. This communication consists of a number of well-defined pathways reminiscent of a complex reaction kinetics scheme (Bottle, 1973). He who requires information must intercept it at one of the several stages it passes through on its path from the original author's mind to where it becomes integrated into the general fund of knowledge. This book's aim is to make the interception of information a much less random process than it normally is. Sometimes this will be achieved via computerized services, but most chemists will supplement these with (or replace them by) the very considerable volume of conventionally printed literature.

Measured in terms of the man-years taken to produce it, the chemical literature is the most expensive tool available to the chemist. He cannot therefore afford to misuse it. About five million papers, reports, patents, etc., have so far been published (and this number is doubling approximately every ten years). Making plausible estimates for the number of man-hours each item represents (and remembering the additional man-years required to abstract it and distil it into storage files such as *Chemical Abstracts, Beilstein, Gmelin,* etc.) and costing this at current levels, the chemical literature probably represents $£10^{10}$–$£10^{11}$ worth of research already completed. Not all papers are of lasting value; many on experimental techniques quickly obsolesce as they are replaced by improved methods. The chemist, therefore, may require data from the past and techniques from the present. The chemist's research can be regarded as processing such information so that questions are thrown up to which answers can be found through experimental work, thus generating new information for the future.

He can make this new information available to the scientific community without undue difficulty and others can then build on it through further work and produce more information — provided that it reaches the right man at the right time. It is in the last stage of this information processing scheme, which we designate research, that the communication links are weakest. Because of the size of the chemical literature, virtually no one now obtains all the information he requires by direct communication. He will get access to most of it through a variety of temporary or permanent stores for the primary sources of information. The lines of scientific communication are indeed tenuous and surrounded by a vast ocean of literature. Nevertheless the means for dealing with this literature problem exist today if only people know how to use them and in the foreseeable future improvements in information storage and retrieval will prevent science from becoming submerged under its own literature. The tendency for information to disperse itself in the literature was recognized as long ago as 1882, when Dr H. C. Bolton addressed the newly formed Chemistry Section of the American Association for the Advancement of Science in these words: 'Chemical literature is characterized by two opposing forces, a tendency to dispersal and an effort to collect the widely scattered fragments' (quoted by Van Patten, 1950). One thinks immediately of the analogous Le Chatelier's Principle. Le Chatelier's Principle, however, is concerned with a system in equilibrium but the above two opposing forces are never long in equilibrium.

Our current literature problems are commonly ascribed to the literature explosion which is often imagined to be a post-war phenomenon. Price (1963) has shown, however, that this is merely the exponential portion of a logistic growth curve which started when science was born in the seventeenth century. Any parameter of science we look at shows this growth pattern — be it number of journals or of abstracts or of entries in *American Men of Science*, the first edition of which (1903) contained about 4000 entries and the twelfth edition (1971–73) about 138 000 entries.* A century ago about 1000 new chemical compounds were discovered per year; now the annual figure is 150 000; and so on: the examples could be continued for several pages yet. Such exponential growth (with a doubling time variously estimated as being between 8 and 20 years) cannot go on indefinitely. In the

---

*The 13th edition of *American Men and Women of Science* (Bowker, 1977) contains only 111 000 entries, so that in North America at least the rate of growth may now be slowing down. The rate of growth does, however, vary with discipline and time, though few statistics will exceed the doubling period of less than one year observed for liquid crystal patents (Bottle and Rees, 1979).

developed countries we are probably approaching a limit where the proportion of scientists in the population is reaching saturation (having regard for the proportion of those above a certain minimum intelligence level and the competitive demands of other professions). In the underdeveloped countries there is no such saturation and indeed we may well see from this source an escalation in the growth of science which will more than compensate for any reduction in its growth rate in the West. It is, however, an interesting thought that some 80 per cent of all the scientists who have ever lived are alive today (Price, 1963).

One consequence of this exponential growth pattern appears to have been largely overlooked. If we take a conservative figure for the doubling time as 15 years, a new graduate will have x units of information available at the start of his career, but when he retires 45 years later the total amount of information available will be $8x$ units. This underlines the necessity to teach students how to use the literature effectively. The main methods used to teach students how to use their subject literature were reviewed by Bottle (1967). They have changed little in the past decade, as can be seen from reviews by Whittington (1976) or Stevenson (1977), despite the recent flood of papers from librarians who have just discovered the importance of user education.

Many of the older generation of chemists were shown at the start of their careers how to use the chemical literature to the best advantage by a senior and more experienced colleague. Today many factors militate against this ideal method of instruction. The need for more formal instruction in this subject results from what can be termed 'the problem of numbers', referring to both the students to be instructed and the amount of material existing, making it difficult for any man who is not a literature specialist to have much expert knowledge outside the field of his particular interest. (This is, of course, why there are so many contributors to this book.) The growth of the technical library and the emergence of the literature specialist, while welcome events in themselves, have tended to place intermediaries between the chemist and one of the most important tools of his trade. While saving him much time and effort in retrieving the information he has requested, this feather-bedding has often resulted in the chemist being insufficiently aware of what treasure actually is hidden, so that he does not take advantage of the facilities which are available. This book aims, not at making every bench chemist his own technical librarian, but to save him and his library colleagues time and enable him to use to the full the excellent scientific library facilities which are within the reach of all in this country today.

The basic problem facing the inexperienced worker is firstly to be aware that the information required may be already in the literature, secondly to be aware that it can be retrieved from storage and thirdly to understand storage and retrieval mechanisms so that all types of information may readily be obtained. There are two main parameters — subject matter and time — which determine which of three main methods of information storage are most appropriate in a particular case. The oldest information storage method is the comprehensive treatise or monograph, in which material is collected from a closely defined subject field over a long time interval. The material is condensed, is often critically evaluated and processed, and is retrievable from storage through the layout of the treatise. The second method involves storing the information article by article as it arrives at the Abstracting Centre, the month's (or week's) catch being broadly classified for current use. Specific information is retrieved from the store through the subject indexes which should preferably be cumulated over a period of years. The third main type derives from the second and its development has doubtless been stimulated by the time lags inherent in the first two types. This type covers all the self-indexing systems, including computer-sorted storage and retrieval systems. Because information is coded into the store, this effectively indexes it and the whole store is continuously and completely searchable.

Recognizing that a knowledge of the structure of chemical literature is the key which unlocks its treasures, the next step is to show how the practical problems connected with the subject are best approached. The first requirement is a clear delineation of precisely what information is being sought. This can sometimes be done by the formal classification of the required information. Classification systems such as the Universal Decimal System are discussed in Chapter 2, and the chemist is well advised to acquaint himself at an early stage with the details of the classification scheme used by his 'home' library for chemistry and cognate subjects and especially those for his special field of interest. There is, however, no set formula for solving all problems. The approach adopted depends on the nature of the problem, the depth of the inquiry and, to a large extent, the experience of the searcher. Use of the imagination can lead to information from unsuspected sources such as government reports and company reports to shareholders (Chapter 15), patent literature (Chapter 14), and even newspapers, company chairmen's statements and the non-literature sources discussed in Chapter 16.

The types of literature query or problem encountered are discussed in some length in Chapter 18. A rough classification is as follows:

(1) *Facts or data related to a specific compound.* This is often the simplest enquiry to make, since the compounds can be easily indexed by their formulae.

(2) *Examples of a specific type of reaction or measurement.* A more

exhaustive search may be necessary than in the previous case, since indexing tends to be of compounds rather than of reactions or measurements. Computer sorting of structural fragments of encoded product and reactant molecules is now an established method in the documentation of organic reactions. (See J. E. Ash and E. Hyde's excellent book on *Chemical Information Systems* (Ellis Horwood/Wiley, 1975), especially Chapters 12 and 14.)

(3) *First footholds in a new field or reconnaissance reading.* This is the problem which faces all workers when they start a new topic or try to keep abreast of recent trends in other fields. It is especially acute for the new graduate when he comes to bridge the gap between undergraduate knowledge and research work. This requirement of background information is dealt with in Chapter 7.

(4) *Correlative material.* The collection, from different areas of chemistry, of facts which are unrelated except for fitting into a particular theory is especially difficult and this is where initiative and imagination can yield the greatest dividends.

(5) *Historical approach.* This is perhaps of greatest interest to the bibliophile. Histories of science and biographical material are discussed in Chapter 17. Skolnik (1976) wrote a succinct history of the subject matter of this book under the title 'Milestones of chemical information science'.

(6) *Industrial development rather than academic research problems.* Apart from information on fundamental chemistry, commercial intelligence is often required as well as a knowledge of the relevant patent literature. (See Chapters 16 and 14.)

(7) *Communication to others.* When the literature is used for this purpose, as in report writing, teaching, and so forth, much is lost if the presentation is ineffective. This is why some guides to this art have been briefly mentioned (Chapter 18). A detailed discussion of this subject is, however, outside the scope of this book.

Types (1) and (2) above represent what might be termed the *everyday approach*, where one requires specific information on a topic in hand. In contrast to this is the *exhaustive approach*, where one needs to make an extensive and detailed survey and where one uses the methods outlined in Chapters 5 and 18. The third approach to the literature is the most time-consuming of them all: this is the *current approach* – i.e. keeping up to date with current progress. Although some pointers on how to tackle this problem are given in the final section of Chapter 18, in one's more despondent moments one feels the solution to this problem must await the invention of the 40 hour day! Some amelioration of this problem can doubtless be achieved through increasing one's reading speed by the methods advocated by E. and M. de Leeuw in *Read Better,*

Read Faster (Penguin, 1969) or by taking a course such as that provided by Carborundum Ltd. The literature is probably most frequently consulted to answer quick reference queries (see Chapter 18). A selection of quick reference queries answered by LADSIRLAC form the basis of Exercise 12 and it will be most instructive for the novice to work through them.

If one understands how the various types of chemical literature arise, one can then locate and evaluate them better. For example, many reports by scientists working in government laboratories are not published in the usual scientific journals. A knowledge of government departments sponsoring such reports or the agencies which collect or catalogue them is essential for searching the literature. Thus the development of the Government's responsibilities for research are traced in Chapter 15, and for similar reasons an outline of the procedure for obtaining a patent is given in Chapter 14. The structure of *Beilstein* must be known if it is to be used when the indexes are not available. This is discussed at length in Chapter 11. (One should not be put off by the fact that *Beilstein* is in German. The language is quite straightforward and can easily be understood by the novice without having to use a dictionary too often.) To help those who do not read German easily, a translation of the contents of Landolt–Börnstein's *Tables* has been appended to Chapter 8.

More attention is paid in this book to English-language sources than to foreign-language ones, since it is thought that the former will usually be more easily obtainable, and also more easily read, by the average chemist now that translation tests for graduating chemists have fallen out of favour. The British aspects of problems of technical information have been discussed more fully than the American aspects, since these are the ones with which the contributors are most familiar; furthermore a recent American guide is available (A. Anthony's *Guide to Basic Information Sources in Chemistry* (Wiley, 1979)). A. Nowak's recent guide *Fachliteratur des Chemikers* (Deutscher Verlag der Wissenschaft, Berlin, GDR, 3rd edn, 1976) is interesting for its descriptions of Russian and German sources.

Having confessed to this sin of nationalistic bias, we may now caution the reader about the existence of this evil in certain handbooks, abstracts, etc., produced in Germany during the Nazi regime; it can also be detected from time to time in Soviet publications. The lack of attention paid by western reviewers to eastern work is a criticism noted in Chapter 7, but one also notices the occasional American article which pays scant attention even to any relevant British work.

One must also remember that every article is a compromise between the ideal of comprehensiveness and the practical need for conciseness. That this latter tendency is increasing is illustrated by comparing the long introductory discourses in papers published half a century ago with the short *pro forma* accounts in today's leading journals. The exponential increase in the literature referred to elsewhere has doubtless been a powerful catalyst to this foreshortening of chemical communication. Some authorities go so far as to propose that journals should publish only abstracts of articles submitted to them, so that interested workers could then write in for photocopies of the original manuscript as and when required, and the Chemical Society has now introduced a synopsis journal (see Chapter 3).

While the primary publication of research papers as abstracts would undoubtedly reduce the physical volume of the literature, it would not help to collect it together but would serve to make it even more difficult for the individual scientist to get at the facts he requires. Similar remarks apply to proposals that all journals should be issued in microform; this has the additional disadvantage that, since it is more trouble to read microforms than to skim through the conventional journal, one will then have to rely even more on a literature-unifying information retrieval service such as *Chemical Abstracts* — even though this imposes an additional delay in getting one's information and cuts down the possibility of browsing in a related field (which can frequently lead to a new line of attack on the problem in hand).

Few people read a technical book for pleasure; rather do they read it for the facts which it contains. One looks up isolated facts for reference purposes and thus reference is the commonest form of technical reading. The object of education is, of course, not to instil isolated facts, since virtually no one can remember all he may ever need, but is rather, to use Whitehead's words, the 'acquisition of the art of the utilization of knowledge'. Thus it is better to learn how these facts may be utilized and correlated and last, but not least, where they may be discovered. To this last objective this book is dedicated.

REFERENCES

Bottle, R.T. (1967). *Progress in Library Science,* Butterworths. pp. 97–115
Bottle, R.T. (1973). *J. Docum.,* **29**, 281
Bottle, R.T. and Rees, M.K. (1979). *J. Inf. Sci.,* **1**, (2), in press
Price, D.J. De Solla (1963). *Little Science, Big Science,* Columbia University Press
Skolnik, H. (1976). *J. Chem. Inf. & Comp. Sci.,* **16**, 187
Stevenson, M. (1977). *J. Docum.,* **33**, 53
Van Patten, N. (1950). *J. Chem. Educ.,* **27**, 431
Whittington, I. (1976). *M.Sc. Thesis,* The City University, London

2

Libraries and their use

F. Earnshaw

Keeping in touch with new developments in science is by no means the simple task that it was in the early part of the century, when the whole body of scientific knowledge was much smaller and less specialized. The rate of scientific discovery has led to a corresponding increase in the literature which announces and surveys these new developments, and it is reported that the world's supply of recorded knowledge is doubling every ten years or less. Libraries are trying to cope with an enormous output of printed material.

Ways of keeping up to date include correspondence with workers in the same field and attendance at scientific conferences, but will also surely involve scanning the literature, and here the scientist must make use of library services. The use of a library is undoubtedly becoming more complex. The apparatus of bibliographical research is in some ways like the apparatus in a laboratory; it cannot be used efficiently without both instruction and practice. Casual browsing, although a potentially profitable method when there is no particular or immediate objective in mind, is dangerous and inefficient in other circumstances. The reader must learn how to use the catalogues, indexes, abstracts and bibliographies which are his guides to the literature of his subject, for only with their aid can he be sure that he has effectively surveyed the field in which he is interested, or that he has discovered the relevant literature in the library he is using.

Readers who wish to pursue the techniques involved in using libraries effectively should consult the works listed at the end of this chapter.

CLASSIFICATION

There is no doubt that every scientist would find it most convenient if all the books he was ever likely to read, whatever their subject, were grouped together in one section of the library. Librarians have a great deal of sympathy with this point of view, but they know that the result would be absolute chaos. The most effective arrangement of books is by subject, but although this sounds simple, it is far from straightforward in practice; the subject classification of books is based on, but very different from, a classification of knowledge, for books are physical objects which, whatever their contents, can occupy only one place in a library. Many books, of course, do not fit neatly into one place in the classification schedules; for example, is a book on the radiochemistry of silicon to be placed with all other books on radiochemistry, or is it to be placed with all other books on silicon in the appropriate section of inorganic chemistry? The linking of two subjects is a concept which can be expressed in some classification schemes, and it is possible where necessary to make entries in the library catalogue under both subjects, but the book can occupy only one place.

It is also inevitable that any division of knowledge into 'subjects' is by and large an arbitrary one, which because it groups together certain topics will also widely separate others. For example, in the Dewey Decimal Classification a study of the subject of metals will necessitate a search of books arranged under the following headings, among others: geology, chemistry, mining engineering, mechanical engineering, strength of materials, structural engineering, metallurgy and manufactures. Similarly, anyone interested in technological developments in a particular country may find that he has to look not only in the section of the library containing books on technology in general, but also in other sections dealing with particular branches of technology, such as medicine, engineering, agriculture and textiles, and also possibly in economics and education.

The general principle on which the librarian works is to classify the book according to its main subject emphasis, putting it where it will be most useful, and to create an index (in the form of the library catalogue) which reveals the decisions made. In this connection the chemist must remember that a book entitled *Mathematics for Chemists* is likely to be considered as a book on mathematics, not on chemistry.

It is increasingly difficult to keep classification schedules in pace with the development of knowledge. The rigid boundary between pure and applied science, which is still maintained in most schemes for the arrangement of books, is becoming more and more unrealistic. In many schemes chemistry is separated from chemical technology, which results, for example, in a wide division between polymer science and plastics technology. Most classification schemes (the Universal Decimal Classification is a good example) have a system of regular amendment and revision by subject experts, which ensures that eventually provision is made for new subjects and new developments. In spite of this the schedules of many schemes have fallen behind modern trends, and new interdisciplinary subjects are particularly difficult to place.

Bearing in mind that librarians, as shown above, are working with imperfect and out-of-date tools, let us now examine some of the more common classification schemes. Perhaps the most widely used system is the Dewey Decimal Classification, which is found in most public libraries and in certain university and college libraries. In this scheme the whole field of human knowledge is divided into nine sections, denoted by arabic numerals 1 to 9, with a tenth section, denoted by the numeral 0, for general works. A three-figure minimum is used, as follows:

000	Generalities	500	Pure sciences
100	Philosophy	600	Technology
200	Religion	700	The arts
300	Social sciences	800	Literature
400	Language	900	General geography and history

Subdivision of subjects is by the use of a decimal point after the third figure, giving nine further divisions at each stage (cf. *Table 2.1*). The classes of most interest to chemists are as follows:

510	Mathematics	590	Zoological sciences
520	Astronomy	600	Technology
530	Physics	610	Medical sciences
540	Chemistry	630	Agriculture
	541 Physical and theoretical chemistry	660	Chemical and related technologies
	542 Laboratories, apparatus, equipment		661 Industrial chemicals
	543 General analysis		662 Explosives, fuels
	544 Qualitative analysis		663 Beverage technology
	545 Quantitative analysis		664 Food technology
	546 Inorganic chemistry		665 Industrial oils, fats, waxes, gases
	547 Organic chemistry		666 Ceramic and allied technologies
	548 Crystallography		667 Cleaning, colour and related technologies
	549 Mineralogy		668 Other organic products
550	Sciences of the earth and other worlds		669 Metallurgy
560	Palaeontology	670–680	Manufactures
570	Life sciences		676 Pulp and paper technology
580	Botanical sciences		677 Textiles
			678 Elastomers and elastomer products

In theory the classification numbers, based on the use of decimals, are capable of infinite expansion. In practice the process of expansion, made necessary in some cases by the introduction of new subjects, leads to many long and unwieldy numbers. Moreover the use of arabic numerals limits the number of main subject classes to ten, and is there-fore directly responsible for some of the long numbers. The scheme was introduced in 1873, and as its main outlines have not been changed, it is now out of line with contemporary thought. In spite of its imperfections the scheme works, however, and many large libraries are so firmly wedded to it that to change to another scheme would be a huge under-taking.

An internationally standardized adaptation of the Dewey system is the Universal Decimal Classification, usually known by its initials as UDC. This scheme is widely used in the USSR, Eastern Europe, the UK and many other countries in government, college and special libraries, and is thought by many librarians to be more appropriate for scientific and technical libraries because of its greater elaboration and flexibility. The main lines of division of Dewey are maintained (although there is now considerable divergence in the sub-divisions), but the three-figure minimum has been abandoned, and additional points are inserted, for convenience, to break long numbers into groups. Although the initial decimal point is omitted, it must be remembered that in reality all numbers are decimal fractions, otherwise the sequence of numbers will not seem logical (see *Table 2.1*).

The differences between UDC and the Dewey classification are growing, and the two schemes are not now identical even up to the third figure. An example is the number for biochemistry: 577.1 in UDC, 574.192 in Dewey. UDC suffers from the same disadvantages as Dewey in the limitation of its main classes to ten, but it is more truly an international scheme and has eliminated some of the Western bias found in Dewey. It is truly international in the arrangements for revision, and is much more radical than Dewey in keeping itself up to date. It can express complex subjects and combinations of subjects by the use of the colon, one of its most useful notational devices:

547	Organic chemistry	547:541.128	Catalysis in organic
541.128	Catalysis		chemistry

Although other (including home-made) classification systems are occasionally encountered, the only other major scheme which chemists are likely to come across is that of the Library of Congress, used mainly in certain university libraries. Here the various fields of know-ledge are designated by capital letters, with a second capital letter to indicate the principal sub-divisions of each main field. The classes of interest to chemists are as follows:

Q	Science	QL	Zoology
QA	Mathematics	QM	Human anatomy
QB	Astronomy	QP	Physiology
QC	Physics	QR	Bacteriology
QD	Chemistry	R	Medicine
QE	Geology	S	Agriculture
QH	Natural history	T	Technology
QK	Botany	TP	Chemical technology

Further subdivision is by arabic numerals, and a combination of two letters and four numbers is common. Long and complex numbers are rare, and on this point the scheme undoubtedly scores over Dewey and UDC. Use of the LC Classification has been fostered by a service of printed catalogue cards, available since 1902. The scheme was individually designed, however, for one large, unique, American library, the Library of Congress, and is not necessarily completely suitable for other libraries. One noticeable feature of the schedules is that they often lack the hierarchical arrangement which is common to both Dewey and UDC; instead a straightforward alphabetical listing is a common device for enumerating the subdivision of a subject. All three systems have the major disadvantage that the pure and applied aspects of chemistry are separated.

Table 2.1   COMPARISON OF THE THREE MAJOR CLASSIFICATION SCHEMES

DDC, UDC or LC subject headings	DDC	UDC	LC
Pure science	500	5	Q
Chemistry	540	54	QD
Physical & theoretical chemistry	541		QD 453–655
Theoretical chemistry		541	
Physical chemistry	541.3	541.1	
Electro- and magnetochemistry	541.37		
Electrochemistry		541.13	QD 552–585
Electrolytic solutions	541.372		
Electrolytic dissociation. Ions		541.132	
Electrolytic dissociation. Hydrogen concentration			QD 561
Hydrogen ion concentration	541.3728	541.132.3	

LIBRARY CATALOGUES AND ARRANGEMENT

In all these schemes, as each volume is added to the library, it is, according to its subject, allocated an appropriate classification symbol; this symbol may be figures, letters or a combination of both, according to the scheme in use, and it is printed on the spine of the book. The books are then arranged in order by the classification symbols.

The pleasure of hoarding for its own sake is reserved for the miser. Storage in a library is for the sake of retrieval, and this is the crux of the librarian's problem. The idea of a librarian merely as a custodian of the contents of his library is now fortunately dead. The accumulation of appropriate material is, of course, one of his essential tasks, but accumulation without easy accessibility would be futile. Accessibility means arranging the material in a logical and systematic way and providing the necessary key to the arrangement by means of catalogues and indexes. The easy and efficient retrieval of any publication placed on the shelves of a library depends largely on the catalogue, which is an index to the collection, and is usually kept on 12.5 X 7.5 cm cards. It is advisable to use the catalogue rather than to go directly to the shelves, as the catalogue represents the complete stock, whereas the books to be found on the shelves at any given moment represent only a portion of the stock, and probably not even the best portion.

The catalogue will reveal what publications are available (1) by a given author and (2) on a given subject. Title entries are not very common, except for distinctive titles such as those of annual review publications. In any case, if the title only of a book is known, entries under the title are to be found in most of the standard bibliographies, such as *British National Bibliography* or *Cumulative Book Index*; by this means the name of the author can be traced, and the reader may then return to the author catalogue to discover whether the book is in the library. In the UK it is usual for the author and subject catalogues to be separate, and for the subject catalogue to be in two parts – a classified catalogue and an alphabetical subject index.

Entries in the author catalogue consist of (1) the main entry for each book, under the name of the author; (2) added entries under subsidiary authors, editors, translators; (3) references from alternative forms of authors' names; (4) entries under the name of important sets or series of works. It is important to remember that the name under which a publication is entered may not be a personal one, but the name of a corporate body which is in certain cases treated as the author. Such a corporate body may be a government department, a learned society, an academic institution or an international conference; if the publication was compiled by an individual, and his name appears on it, there will be an added entry under his name. The object of the cataloguer is to ensure that from whatever angle the user approaches the author catalogue in his search for a particular work, he will find an entry for it, or will be guided to the main entry.

In some cases, this part of the catalogue may contain entries under the names of persons written about, in the case of biographies or works of criticism. It is then usually known as the name catalogue rather than the author catalogue.

The card (*Figure 2.1*) for each main entry (and in some libraries the added entries also) will give the classification symbol of the publication, so that the enquirer will be directed to the correct place on the shelves.

HEYS, Harry Law

Physical chemistry.   5th ed. London,
Harrap, 1975.
512p. ill. 22 cm.

541.3

Figure 2.1

In the classified catalogue the arrangement of the cards corresponds to the arrangement of the books themselves on the shelves of the library. The prominent feature on each card is therefore not the author's name but the classification symbol, and the cards are arranged by these symbols. This gives an arrangement by subjects and their sub-divisions (*Figure 2.2*).

541.3

HEYS, Harry Law

Physical chemistry.   5th ed. London,
Harrap, 1975.
512p. ill. 22 cm.

Figure 2.2

The alphabetical subject index consists of cards arranged alphabetically by the names of subjects, giving for each subject its classification symbol. Thus, by looking up Chemistry in this index, it will be found that the appropriate classification symbol in UDC is 54. This leads the enquirer to the correct place in the classified catalogue, where entries will be found for all books on Chemistry in the library.

An alternative form of subject catalogue, more popular in the USA than in the UK, is the dictionary catalogue. Here the cards are arranged in alphabetical order of subjects, and entries are made under the most specific subject heading possible. An elaborate system of cross-references is essential, and references to related subjects are inevitably scattered. As the catalogue itself is arranged in alphabetical order by subjects, no alphabetical subject index is necessary. It is possible to inter-file the cards with those of the author catalogue, so that the whole catalogue is in one complete alphabetical sequence. *Figure 2.3* is an example of a subject entry for a book in a dictionary catalogue.

PHYSICAL CHEMISTRY

HEYS, Harry Law

Physical chemistry.   5th ed. London,
Harrap, 1975.
512p. ill. 22 cm.

541.3

Figure 2.3

Periodicals may be arranged in classified order, or alphabetically by title, or in numerical order with each title receiving a code number. They are often in a special room or section of the library, and bound volumes are usually filed separately from current issues. Sometimes the periodicals held by a library are entered by title in the author catalogue; sometimes they are in a separate list or index. Where part of a library's holdings are in microtext form (microfilm, microcard or microfiche), special reading apparatus is necessary, and enquiries should be made at the library counter.

Pamphlets are often stored separately from books, in special pamphlet boxes or in vertical filing cabinets. If they contain information of permanent value, they are classified and catalogued in the usual way, but those which are ephemeral may be collected in broad subject groups, with no specific catalogue entry.

Patent specifications and standard specifications are usually filed in separate sequences by their serial numbers, and are not normally entered individually in the catalogue. Practice may, however, vary, and it is best to consult the librarian.

Cataloguing practice varies also for government publications, trade literature and theses, and here again the guidance of the librarian should be sought.

Although the main book stock is normally arranged in numerical sequence by classification symbols, it is as well to realize that the sequence may be broken or changed to suit the particular needs of individual libraries. Older material may be kept in the stack, to which access may be restricted. Oversize books may be in a separate sequence. There may be special collections which are kept together as a unit, although the books in them cover many subjects. Because of these variations, many libraries display a plan showing the exact location of the various subjects and types of literature, and this, coupled with adequate guides on the bookshelves, should make it easy for users of the library to find what they require. In some libraries a booklet describing the library, its arrangement and its services is available on request. In any case the librarians are normally only too glad to help any reader in his search for appropriate information; indeed this is their primary task, if they are to regard themselves as more than mere custodians.

MAJOR SCIENCE LIBRARIES

Space does not permit the inclusion of all important libraries, and perhaps there is little need to describe here such well-known collections as those of the British Library in London; the Bibliothèque Nationale in Paris; the Lenin Library in Moscow; or the Library of Congress in Washington.

Special mention must be made of the Science Reference Library (previously the National Reference Library of Science and Invention) in London, which is now part of the Reference Division of the new British Library. Eventually it will be rehoused in a new building in the Somers Town area of London, north of Euston Road, along with the British Museum Library, but at present it is separately housed. The nucleus of its collection came from the former Patent Office Library, but its scope and depth of coverage have been greatly extended. The aim is to make a comprehensive collection of periodicals in any language and books in

English, German and French. Books in other languages are taken selectively. Patent literature is taken comprehensively.

The Science Reference Library is at present organized into two complementary units. The Holborn Branch, in Southampton Buildings, off Chancery Lane, London WC2, is the former Patent Office Library. The Bayswater Branch, housed over Whiteley's department store in Porchester Gardens, London W2, holds all the additional literature (especially biological) resulting from the widened scope. The Library is open to all comers without formality and a printed guide is available free on request. As a reference library it does not, of course, lend any of its stock — with the advantage that readers are unlikely to find that the items wanted are out, unless they are being bound. To make its resources widely available it operates a swift photocopy service on either a postal or a while-you-wait basis. The scientific staff are ready to help readers exploit the stock and will also handle inquiries received by post, telex or telephone. A range of catalogues and indexes is available in each Branch, including one of bibliographies and another of glossaries and dictionaries. Also valuable is the linguistic aid service, in which one of the library linguists is prepared (by appointment) to sit with the scientist reader to help him understand the gist of an item and thereby evaluate its relevance. The Science Reference Library is the library to consider whenever local resources fail to provide the answers required. A special visit of an hour or two during a visit to London can be rewarding, but it is advisable, particularly if time is likely to be short, to get in touch with the Library in advance to get guidance on which Branch is likely to hold most of the material sought. On the other hand, if local resources have produced the references but not the original articles, then photocopies may be obtained from the Science Reference Library or the British Library Lending Division.

The need for a country-wide lending service is met by the British Library Lending Division (BLLD), which is located at Boston Spa, Yorkshire. This library, formed by the amalgamation of the National Lending Library for Science and Technology and the National Central Library, has a very large stock covering all subject fields, though it is particularly strong in science and technology. In March 1978 49 300 serial titles were currently received, and a further 88 000 non-current titles were also held. The stock also included 1.7 million monographs on all subjects, and a very large collection of scientific and technical reports. Special efforts are made to collect conference proceedings in all subjects and translations into English from all languages. The bulk of the collection of translations is from the Russian language. Direct borrowing facilities are available to other libraries; the individual should enquire first through his own library, unless he lives near enough to Boston Spa to use the reading room there.

Although the Lending Division of the British Library plays the major role in supplying literature not available locally, many libraries throughout the country lend to each other as part of a national system of inter-library loans which has existed for many years. Books and periodicals may be borrowed from abroad when necessary. Public, university, college and many special and government libraries play their part in this scheme, and the scientist should enquire at his own library or at his local public library.

Mention should also be made of the Chemical Society's Library in Burlington House, Piccadilly, London W1. Although somewhat starved of recent material owing to the ravages of inflation, a postal borrowing and photocopying service is available to members. It has a useful collection of older material.

Aslib (3 Belgrave Square, London SW1), formed by the amalgamation of the Association of Special Libraries and Information Bureaux with the British Society for International Bibliography, plays an invaluable part in directing enquirers to the correct source of information, and has an active research and consultancy programme. Membership of Aslib includes libraries of all kinds, government departments, professional institutions, colleges and universities, and industrial and commercial concerns, in the UK and overseas. Publications are borrowed if necessary for members. Aslib has published many bibliographical aids, maintains the Commonwealth Index of Unpublished Translations and can provide information on specialist translators (see Chapter 4). Within Aslib there are 'subject groups', including a chemical group, which have the aim of promoting close contacts and co-operation between workers in similar fields, and of improving information services within the industry.

PUBLICATION DETAILS OF BOOKS

One type of enquiry which librarians meet frequently is for assistance in tracing publication details of books. Many libraries have a special section of general and subject bibliographies; in other cases subject bibliographies are shelved with books on the same subject. For current British publications on all subjects the *British National Bibliography*, published since 1950, is the most useful guide, and gives full details each week of newly published books, in a classified subject arrangement, with author and title indexes. It cumulates eventually into an annual volume, and finally into multi-year cumulations. Many countries have a similar national bibliography.

The American *Cumulative Book Index* attempts to include all

English-language books irrespective of country of publication. The arrangement here is of authors, titles and subjects in one alphabetical sequence.

For tracing details of books which are still in print and which can therefore still be purchased in a bookshop, irrespective of date of publication, *British Books in Print* is the best source of information for British books. For US publications the annual *Books in Print* and *Subject Guide to Books in Print* are the standard bibliographies.

Many good booksellers help their customers to keep in touch with new publications by issuing lists or sets of cards.

New Books in Chemistry is a fortnightly bulletin from Chemical Abstracts Service in their *CA Selects* series, which commenced in 1977.

Some bibliographies help in tracing books in a rather more restricted field. Those most useful to chemists are listed at the end of this chapter (see also page 113).

NEW DEVELOPMENTS

The rapid growth of scientific information has begun to pose serious problems for both scientists and librarians. More and more attention is therefore being given to improving means of exploiting the information stored in libraries. Now, at last, growing numbers of scientists are beginning to realize that their skill and knowledge must be applied not only to creating information, but also to controlling it, and librarians and scientists are working together on this problem.

Modern technological developments, particularly in the fields of computers, microforms and communications, will eventually alter profoundly the traditional operations of libraries. Computer-based current awareness and retrospective searching systems are already well established. Computerized storage and retrieval of text will probably be confined initially to specialized highly used scientific data, made available through national data centres. The conversion of ordinary texts to machine-readable form is not expected for some time, and it seems clear that books in the codex form will continue to be the main vehicle for the storage of textual information into the foreseeable future.

Microform technology and facsimile transmission are the other two most significant developments, and both will eventually play an important role in the storage and communication of information. Growing international co-operation will help to ensure that libraries continue to play a significant part in the task of providing scientists with the information which they need.

SELECT BIBLIOGRAPHY

Using libraries

Burrell, T.W. (1969). *Learn to Use Books and Libraries*, Bingley
Carey, R.J.P. (1966). *Finding and Using Technical Information*, Arnold
Chandler, G. (1974). *How to Find Out*, 4th edn, Pergamon Press
Downs, R.B. and Keller, C.D. (1975). *How to Do Library Research*, 2nd edn, University of Illinois Press
Gates, J.K. (1974). *Guide to the Use of Books and Libraries*, 3rd edn, McGraw-Hill
Whittaker, K. (1972). *Using Libraries*, 3rd edn, Deutsch

Classification schemes

Dewey Decimal Classification and Relative Index (1979). 3 vols, 19th edn, Albany, NY, Forest Press. Many libraries still use the 18th edition (1971)
Universal Decimal Classification. Complete English edn, 4th international edn. British Standards Institution, 1943 – in progress (publication has not yet been completed, but an abridged edition of the whole schedule is available)
United States. Library of Congress. *Classification*. Washington, Government Printing Office, 1901– (issued in sections, each section revised and reissued separately)

Libraries – international

Lewanski, R.C., Ed. (1976). *Subject Collections in European Libraries*, 2nd edn, Bowker
Internationales Bibliotheks-Handbuch (World Guide to Libraries) (1974). 2 vols, 4th edn, Verlag Dokumentation, Munich; Bowker, New York
Steele, C.R. (1976). *Major Libraries of the World; a Selective Guide*, Bowker
The World of Learning: Directory of the world's universities, colleges, learned societies, libraries, museums, art galleries and research institutes, Europa Publications (annually)

Libraries – Great Britain

Aslib Directory. Vol. 1: *Information Sources in Science, Technology and Commerce*. E.M. Codlin, Ed. (1977). 4th edn, Aslib
Esdaile, A.J.K. (1947). *The British Museum Library*, 2nd edn, Allen and Unwin
Irwin, R. and Staveley, R., Eds. (1961). *The Libraries of London*, 2nd revised edn, Library Association
Saunders, W.L., Ed. (1976). *British Librarianship Today*, Library Association. Five chapters are devoted to the British Library

Libraries – United States

American Library Directory, Bowker (a guide to public, academic and special libraries in the USA and Canada, indicating number of volumes and any special subject interests, revised biennially)

Ash, L., Ed. (1978). *Subject Collections: a guide to special book collections and subject emphases as reported by university, college, public, museum and special libraries in the United States and Canada,* 5th edn, Bowker (more than 45 000 entries of collections, including size and specialization)

Young, M.L., Kruzas, A.T. and Young, H.C., Eds. (1977). *Directory of Special Libraries and Information Centers,* 4th edn, Gale Research (entries for 13 000 specialized libraries in the USA and Canada, including not only special libraries and information centers, but also major special collections in university and public libraries)

Bibliographies

Aslib Book List: a monthly list of selected books in the fields of science, technology, medicine and the social sciences, Aslib, 1935–

Technical Book Review Index, Special Libraries Association, New York, 1935–

List of Accessions to the Library, Science Museum Library, London, 1931–

Science Books: a quarterly review, American Association for the Advancement of Science, 1965–

3
Primary sources

C. R. Burman

In the early days of science the dissemination of information was mainly a matter of personal communication either orally or by correspondence. By the end of the seventeenth century many scientific societies had been formed at which members delivered papers to each other describing investigations they had carried out. It was as a natural development from this that they began to publish journals to report the papers delivered and the discussions which followed and so make these discoveries known to absent members and other interested scientists. One of the first of these journals was the *Philosophical Transactions of the Royal Society*, which commenced publication in 1665. By the beginning of the nineteenth century there were about 100 scientific journals in circulation, the number increasing about tenfold by the middle of the century and reaching approximately 5000 by 1900 (Cook, 1966).

Many attempts have been made to estimate the total number of scientific periodicals currently being published. Diana Barr (1972) quotes estimates that in 1965 there were 113 000 titles in all subjects, and 26 000 worthwhile titles in science and technology in 1967, the latter figure being based on the intake into the British Library Lending Division at Boston Spa. The BLLD's catalogue of *Current Serials Received*, published in 1976, however, listed about 46 500 titles, excluding discontinuations and superseded titles, received in September 1975 from over 100 countries, including 3000 Cyrillic titles, with an additional 4000 titles reported as being on order. Part of the problem in estimating current totals is the difficulty in defining criteria to determine whether a publication is strictly a periodical and whether it has any scientific value, which would exclude mere annual reports. The question of changes of title, mergers, etc., complicates it still further.

Originally the only primary source of scientific information, journals eventually began to include abstracts of articles appearing in other, particularly foreign, periodicals, thus adding a current awareness service to their original archival function — e.g. the *Journal of the Chemical Society* from 1853. This function became steadily more important as the science of chemistry developed and the professionalism of its practitioners increased and as their growing numbers eventually reduced the possibility of personal attendance at centralized scientific meetings.

It has been estimated (Wootton, 1976) that current scientific periodicals increased numerically by 4% per annum until the late 1960s, and that between 1960 and 1970 they increased in size by 50%. Commercial publications increased their average number of pages from 500 in 1969 to 1041 in 1973; society publications rose merely to 622 in 1973, as against the same base line of 500 in 1960.

Between 1970 and 1975 the number of current journals monitored by *Chemical Abstracts* rose from 9294 to 11953, although the average number of articles remained constant at 34 per annum per journal. The total number of documents cited by *CA* in 1975 was 454300 (a 21% increase over 1974) but the relative percentage of journal articles declined from 75% of documents abstracted in 1970 to 69% in 1975. An increase, however, was recorded in the number of conference proceedings and similar collections of papers from 4603 in 1970 to 7253 in 1975 and in the number of papers delivered per conference from 10.2 in 1970 to 15.6 in 1975 (Baker, 1976).

In spite of the growth of primary scientific publications such as patents, conferences, reports and theses, periodicals still remain the prime method for communicating chemical and chemical engineering information. Their importance lies in their convenient physical format and regular and frequent publication which permits the inclusion of the latest information. This information can with careful editing and refereeing maintain a very high standard and deal with highly specialized topics at greater or shorter length and detail according to their significance. If properly indexed, journals fulfil an archival function by providing a permanent and retrievable record of work done. The varied nature of their contents provides an element of serendipity and browsability which cannot be evaluated.

Journals are often used, by means of letters to the editor, to establish priority in a certain field of research, to inform colleagues of new work being undertaken or to indicate the stage of development which has been reached. Periodical articles often include much information that cannot be abstracted, such as advertisements, rejected methods and references to other work quoted by authors, and so a chemist should not depend entirely on abstracts for his reading.

TYPES OF PERIODICALS

Scientific journals can be grouped according to the type of information they contain and the nature of their publishers, though these categories frequently overlap. The most important journals from a scientific point of view are those dealing with original work, most of which are published by learned societies and similar institutions, such as universities and research associations. Others, issued by professional bodies and trade associations, usually emphasize the manufacturing and technical aspects of industry, while an increasing number of journals being published by commercial publishers stress commercial and financial developments. This latter type of journal usually depends on revenue from advertising and subscriptions. A further type of journal sometimes overlooked is the house journal published by industrial firms, some of which are noted for their high scientific content and others for their promotional features. These are, however, discussed along with other types of trade literature in Chapter 15.

Periodicals reporting original research

Society publications

Though the learned societies were the first publishers of scientific journals to further their particular scientific interests, their publications are still the most important vehicles for the dissemination of basic research data. Their wide distribution and careful scrutiny by expert referees lend them great prestige and enable them to achieve a consistently high level. Societies maintain ownership and control of their journals in a variety of ways (Anon., 1963).

The *Journal of the Chemical Society*, itself the oldest national chemical society journal in the world, contains original work communicated to the Society or sponsored by its Fellows. Commencing in 1841 as the *Proceedings of the Chemical Society*, it merged in 1849 with the *Memoirs and Proceedings* (1841–48) to form the *Quarterly Journal of the Chemical Society*, and then appeared under the title *Journal of the Chemical Society* from 1862 to 1965. From 1966 to 1971 it was published in three sections – A: Inorganic, physical, theoretical, B: Physical, organic, and C: Organic. Since 1972 it has been divided into 6 publications:

(1) *Journal of the Chemical Society, Chemical Communications*, which is the continuation of *Chemical Communications* and provides a medium for the publication of brief papers in all branches of chemistry containing urgent novel results.

(2) *Journal of the Chemical Society, Dalton Transactions*, the successor in part of *JCS, Section A*, and the organ for papers on the structure and reaction of inorganic compounds and applications of techniques of physical chemistry.

(3) *Journal of the Chemical Society, Faraday Transactions I*, which contains papers on physical chemistry and

(4) *Journal of the Chemical Society, Faraday Transactions II*, the section for chemical physics.

These two sections replace the *Transactions of the Faraday Society* and *JCS, Section A*.

(5) *Journal of the Chemical Society, Perkin Transactions I*, which replaces *JCS, Section C* and contains papers on organic and bio-organic chemistry.

(6) *Journal of the Chemical Society, Perkin Transactions II*, which is the physical organic section replacing *JCS, Section B*.

Between 1871 and 1926 *JCS* contained abstracts on pure and theoretical chemistry for which author and subject indexes were provided. Separate indexes are also available to the scientific communications published in the *Proceedings*. The latter began to be issued separately in 1957 to record monthly news of the Society and its activities, reports of foreign congresses and short communications of scientific importance. Since the beginning of 1965 it has been replaced by *Chemistry in Britain*, a joint publication of the Chemical Society and the Royal Institute of Chemistry, whose *Journal* it also incorporates. The Chemical Society also publishes *Education in Chemistry*.

The Society of Chemical Industry is another British institution with several important primary publications. The *Journal of Applied Chemistry and Biotechnology* is a monthly publication of original papers and reviews in specialized fields with a large but selective abstracts section. The delay in publishing indexes limits the latter's usefulness. Another SCI publication is the monthly *Journal of the Science of Food and Agriculture*, containing original work as well as important review articles and an extensive abstracts section to which an author index is issued.

Other British societies publishing journals containing original work include the Faraday Society (now a division of the Chemical Society) (*Faraday Symposia* and *Faraday Discussions*); the Biochemical Society, which publishes the *Biochemical Journal*; the Institution of Chemical Engineers (*Transactions* and *Chemical Engineer*, which contains news and articles of general interest); the Institute of Petroleum (*Journal*); the Oil and Colour Chemists' Association (*Journal*); the Pharmaceutical Society, which issues the weekly *Pharmaceutical Journal* and the monthly *Journal of Pharmacy and Pharmacology*; and the Society of Glass Technology, which publishes *Glass Technology* and *Physics and Chemistry of Glasses*.

The American Chemical Society, which does work similar to that of several British societies, is responsible for a large number of varied publications. The *Journal of the American Chemical Society* dates from 1879 and is one of the most important journals in the world, publishing original work in all fields of chemistry. In addition to original papers, it contains short communications to the editor and expert book reviews. The *Journal of Physical Chemistry* has been published annually since 1896 and contains many important papers and symposia in its field. The *Journal of Organic Chemistry*, started in 1936 because of the growth of the organic side of chemistry, publishes original papers on the theoretical and practical aspects of the subject from a comprehensive or critical point of view. Since 1962 it has been counterbalanced by the journal *Inorganic Chemistry*, which contains original articles on a wide range of topics on the synthesis and properties of compounds and new data, in addition to sections devoted to notes, letters to the Editor and book reviews. In 1968 the ACS began publication of *Macromolecules*. In order to keep industrial chemists and chemical engineers supplied with authoritative data as soon as possible, the ACS also publishes *Chemical Technology*, a monthly publication which replaces the earlier *Industrial and Engineering Chemistry*.

Another ACS publication is the *Journal of Medicinal Chemistry*, which accepts papers displaying the relationship of chemistry to biology, while *Biochemistry* (1962–) caters for original research in fundamental biochemistry, with emphasis on the related contributions of chemistry to biochemistry and of biochemistry to other biological sciences.

In the field of documentation the ACS quarterly *Journal of Chemical Information and Computer Sciences* publishes papers on all aspects of the chemical literature. The *Journal of Chemical Education* covers many topics of interest to teachers of chemistry in articles written in a clear and accurate fashion for the informed non-specialist. Features include descriptions and listings of new products and new literature, while its advertisements and book reviews are most valuable adjuncts.

The journals of the American Chemical Society also appear in microfilm editions simultaneously with the hard copy and are leased to organizations under terms which allow them to make photocopies for distribution without further formality.

Examples of journals of the chemical societies of other countries which carry original work are *Acta Chimica Scandinavica*, in two series, published jointly by the Chemical Societies in Denmark, Finland, Norway and Sweden; *Bulletin of the Chemical Society of Japan*, which publishes papers in English, French and German, and also contains a contents list in English of the papers published in *Nippon Kagaku Kaishi* (Journal of the Chemical Society of Japan, Chemistry and Industrial Chemistry); and the Royal Dutch Academy of Sciences' *Chemica Scripta*.

Some research organizations publish journals. One example is the *Australian Journal of Chemistry*, published by the Commonwealth Scientific and Industrial Research Organization. It contains many papers by CSIRO staff and by university research workers of Australia and other countries. Another example is the *Journal of Research of the NBS* (*see* Chapter 8).

A small number of periodicals are issued by international bodies. One is *Analytica Chimica Acta*, published monthly on behalf of the Section of Analytical Chemistry of the International Union of Pure and Applied Chemistry (IUPAC). Another is *Pure and Applied Chemistry*, which is the official journal of IUPAC as a whole. It is published at irregular intervals and aims to act as a centre for reports and articles prepared by all the various commissions, sections and divisions of IUPAC. Many of the lengthy papers the journal contains have been issued as separate publications, particularly the proceedings of symposia.

Commercial publications

In general, the journals issued by the learned societies deal principally with basic research, though there are some written from a professional or commercial standpoint. However, the learned societies are by no means the only publishers of journals concerned with pure research, and a growing number of important periodicals which the learned societies would be unable to support is being produced by commercial publishers. One of the most influential of these is *Nature*, published by Macmillan since 1869.

Many primary journals of chemical interest are published by Pergamon Press. Among the most important of these are the *Journal of Inorganic and Nuclear Chemistry*, established in 1955; *Talanta*, which publishes original papers, short communications and critical reviews on all branches of analytical chemistry not already covered by the specialist journals; and *Tetrahedron*, which covers all aspects of organic chemistry. Another major publisher of scientific journals is Elsevier, whose publications include the *Journal of Chromatography*, the *Journal of Electroanalytical Chemistry* and *Biochimica et Biophysica Acta*.

Several important journals are published by Wiley-Interscience. The *Journal of Applied Polymer Science* complements the more theoretical *Journal of Polymer Science*, which appears in parts A-1, A-2, B, C and D. The *Microchemical Journal* discusses original investigations and published literature on chemical experimentation involving small quantities of materials.

In Germany many leading chemical journals are published commercially. Verlag Chemie publish *Justus Liebigs Annalen der Chemie*, covering research in all branches of chemistry with a traditional bio-chemical slant. It comprises approximately 2300 pages per year. The fortnightly *Angewandte Chemie*, from the same publisher, reports and reviews developments of current importance, and preliminary communications and conferences, and contains excellent literature reviews. An English-language version is also published. They also issue *Chemie-Ingenieur-Technik*, which describes original work on processes, apparatus, materials and methods of interest to the chemical industry.

Industrial and professional periodicals

Much useful technical information can be found in journals published by trade and professional associations, and industrial, financial and commercial news in those published by commercial publishers. Together they contain a wide variety of features. Very often their advertisements provide useful information on products and processes which have been developed and have become commercially available. Sometimes these journals publish special issues dealing with specific topics of current interest, pinpointing new developments and describing the current technical situation. Trade journals, in particular, often feature translations of papers of interest which have appeared in the foreign press or give general reviews covering broad topics culled from similar journals. Abstracts of articles from domestic and foreign periodicals of interest to a particular industry often feature in their columns, which occasionally list new relevant patents taken from the patents journal or record new trade names, trade marks and companies appropriate to the journal's readers. Supplements are often published in the form of year-books and buyers' guides which indicate specialized sources of supply of equipment and products, together with much other useful data: examples are the *Chemical Industry Directory & Who's Who, Chemistry & Industry Buyers' Guide, Chemical Week Buyers' Guide* and the *Chemical Manufacturer & Aerosol News*.

Lists are published of new trade literature and catalogues which have been received in the editorial office, and this is frequently the only source of such information. Abstracts of trade catalogues are found, in particular, in the journals *Chemistry & Industry, Chemical Engineer* and *Chemical Engineering*.

Details of prices, production figures and market reports are often quoted with other items of commercial and financial importance. A new information service to management in the chemical industry was started by the American Chemical Society in 1973. *Chemical Industry Notes* is a weekly digest of articles selected from over 40 major trade periodicals and newspapers of the main industrial countries, dealing with marketing, investment and production in all branches of chemical industry. Each issue contains a keyword subject index and a listing of state of the art reviews from the previous week's *Chemical Abstracts*. Prices for several hundred chemicals in Belgium, France, Germany, Holland, Italy and the UK appear in the market report published weekly by *European Chemical News* together with US prices taken from the *Chemical Marketing Reporter*.

Book reviews are chosen for their particular relevance to the journal's readership and frequently feature material published by societies, government bodies and industrial firms, which is often difficult to trace through the standard bibliographies. Social and personal news, appointments and obituaries, other general items likely to be of interest to chemists in industry and the academic world, and news of conferences, meetings and symposia are often reported. Where such items are properly indexed, the journal later becomes a useful source for tracing much information of a general nature. Many journals also maintain their own information bureaux, to which subscribers may apply for information, especially from advertisers, by means of prepaid postcards. Examples of the better-known commercial journals are *Chemistry & Industry, Chemical Age, Chemical Week, European Chemical News, Chemical and Engineering News, Chemische Industrie, Europa Chemie, Chemische Industrie International, Chemical Marketing Reporter* and *Japan Chemical Week*.

Examples of journals on applied industrial technology are *Food Technology, British Plastics and Rubber, International Flavours and Food Additives* and the *Paper Trade Journal*. Many journals of this type are published by McGraw-Hill and the Tothill Press.

Specialized periodicals

The earliest chemical periodicals were general in nature, but as chemical science developed, journals began to be established in specific branches. These, like the *Zeitschrift für analytische Chemie*, founded by Fresenius in 1862, were at first limited to the technical side. Gradually others followed, and today each major country has at least one journal in each main branch of chemistry and in many cases in each of its sub-branches. Many of these were founded by groups of specialist chemists and have become the principal journals in their field, such as the *Journal of Chromatography*, the *Journal of Applied Chemistry and Biotechnology*, the Journal of the American Institute of Chemical Engineers (*A.I.Ch.E. Journal*) and *The Analyst*, which was the organ of the Society for Analytical Chemistry but is now the analytical journal of the Chemical Society.

As specialization grew, organizations of industrial chemists began to found periodicals devoted to individual industries. Some of them, such as the *Journal of the American Leather Chemists' Association* and the *Journal of the Society of Dyers and Colourists*, contain important original research, and many of them have useful abstracts sections.

In addition to specialization by subject, specialization by form has also taken place, leading, for instance, to the publication of journals devoted exclusively to abstracts or to reviews. Because information appears in an abbreviated form and for full details one must look at the primary publication, these are called secondary publications and discussed in Chapters 5 and 7.

The very earliest journals, such as *Philosophical Transactions*, carried reports of work carried out in other countries and recorded in other journals, and there are now several hundred abstracting, indexing and alerting journals whose function is to give abridged accounts of items appearing in the world literature. The American Chemical Society, for example, publishes a semi-monthly *Single Article Announcement* which reproduces the tables of contents of about 18 primary journals in the ACS system from which subscribers can purchase specific articles, and *Abstracts of Papers*, which contains abridgements of papers to be presented to ACS divisions and technical sessions at forthcoming national meetings.

Abstracts also feature among the contents of many basic journals, such as *Angewandte Chemie*, which contains abstracts of selected progress reports and review articles, and *Analytische Chemie*, which covers general analytical chemistry, particularly products and fields of application, and biochemical and clinical analysis. Chemical engineering papers from about 500 journals are referenced monthly in *Current Chemical Engineering Papers*, published for the European Federation of Chemical Engineering, while Highlights from Current Literature is a regular feature in *Chemistry & Industry*. *International Chemical Engineering*, published quarterly by the A.I.Ch.E., contains selective translations of current world literature in foreign languages, while the *Journal of the American Oil Chemists Society* also contains a substantial abstracts section.

In addition, there are the hundreds of specialized journals devoted solely to abstracts, which are considered in detail in Chapter 5.

Owing to the delay in publication of the basic journals, many special publications have begun in recent years which aim at rapid publication of urgent papers. *Journal of the Chemical Society, Chemical Communications* was started in 1965 to publish urgent preliminary accounts of new work. *Tetrahedron Letters* claims publication within four weeks of new or urgent work on organic chemistry, and *Inorganic and Nuclear Chemistry Letters* is issued for the same purpose as a supplement to the

Journal. Nevertheless Kean and Ronayne (1972) showed that fewer than 30% of the papers in *JCS, Chemical Communications* and *Tetrahedron Letters* are followed by a full report within two to three years and that, in general, they revealed little of the urgency which their publication would suggest. In addition, a new type of periodical publication has been developed to provide information in advance of the abstracting journals, consisting in many cases of lists of titles which can be produced very rapidly and scanned easily. *Chemical Titles, Current Contents* and other examples of such alerting services are dealt with in Chapter 5. (See also Haglind and Maizell, 1965.)

ALTERNATIVES TO THE CONVENTIONAL PERIODICAL

The growing number of periodicals in recent years has led to their increasing failure to fulfil the basic function of a scientific journal, which is to provide a record of work done that is properly referenced and indexed for later retrieval while at the same time keeping its readers informed in good time of important new research currently being carried out in their field. The increase in the number of titles and the growing specialization of journals have resulted in relevant matter being scattered through a plethora of titles, and the dispersion of this material through individual volumes, which have increased in size with the number and length of articles, has wasted much of the precious time that a chemist can devote to the literature. When the normal delay in publication and the increasing costs of production and distribution are added to these problems, it is not surprising that interested parties have been led to consider how to isolate the essential data required by chemists from the mass of information found in articles and make them available as quickly as possible to the specialists particularly interested, without depriving others of the possibility of consulting them. These considerations have produced various proposals for reducing the cost and increasing the flexibility of journals by publishing in non-conventional form, mainly involving the use of microforms.

The journal *Chemie-Ingenieur-Technik* for some time has appeared in a new format, with technical articles that were formerly highly specialized and rather lengthy now consisting of single-page summaries only. The complete articles are available in microform on application.

The American Chemical Society has been publishing its journal in microform since 1968 and offering microfiche versions of supplementary material such as tables, computer printout, etc., not included with the hard copy edition. Recently experiments have been conducted of printing sections from the *Journal of Organic Chemistry* and the *Journal of Physical Chemistry* in miniprint, readable only by means of a hand lens, to see if supplementary material in this format could be included in the same issue. The American Chemical Society has also researched the possibility of producing *JACS* in two formats simultaneously, one comprising short articles of key information as a summary journal for individual readers, and the other for library archival use consisting either of short typeset articles reproduced from camera ready copy supplied by the author with full supporting material in miniprint as a supplement, or the complete article in miniprint.

A similar two-part system of publication is being tried by *European Chemical Reports* with typescripts of complete articles being available in microform simultaneously with short synopses of specialist interest.

After distributing samples the Chemical Society started publication of a synopsis journal in 1978. This covers organic, inorganic, physical and analytical chemistry, using illustrative matter where possible and with synopses specially written by the authors for this purpose and including the main references. The full-text version is available in two formats: a negative diazo microfiche version produced from typescript in which references are repeated at the foot of each page with structural formulae repeated so as to bring them near to the relevant text, or, as an alternative, a miniprint edition of the full text in a 3:1 linear reduction. It is intended that the new monthly journal should be additional to and not a substitute for *JCS*. While the full text in the microform version can be in English, French or German, synopses are in English.

The Chemical Society has also co-operated with the Gesellschaft Deutscher Chemiker and the Société Chimique de France in designing a primary publication, the *Journal of Chemical Research* (1977–). This is also intended to be a general journal in two parts, one part containing synopses of one or two pages in English, and the other a microfiche or miniprint version in French, German or English.

In the Soviet Union VINITI (the All-Union Institute of Scientific and Technical Information) accepts edited author manuscripts as an alternative to publication in journal form and announces them in *Katalog Deponirovannykh Rukopisei* and in *Referativnyi Zhurnal*. Copies can be purchased in microform or on paper.

Senders (1977) has recently described an on-line journal where papers are stored in a computer network and read on a VDU when requested.

INDEXES TO PERIODICALS

Keys to the contents of periodicals are provided by both abstracts and indexes: abstracts are intended mainly to keep their readers informed of recent progress, while the chief purpose of an index is to assist in retrospective searches.

Indexes can take many forms, one of the commonest being the general index to a single volume. Most periodicals issue an index to each volume, usually annually, and this is normally arranged in alphabetical order, though there may be separate indexes for authors and subjects. The subject index may be under specific headings or may consist of group entries. The author index, which seems simple enough, may contain pitfalls due to the use of different systems of transliteration; the presence or omission of umlauts in German names or their replacement by the letter 'e', which will affect the alphabetical sequence; prefixes such as 'de', 'de la', 'du', 'von', 'van der', which sometimes precede and sometimes follow the surname; hyphenated names, which may be entered under the first or second part of the name; the arrangement of names beginning with 'Mac', 'Mc', 'St'; the occurrence of names with variant spellings; and even the method of alphabetization itself.

In general, as this type of index covers only one year of one particular journal, its use for literature searching is limited. Sometimes, however, such indexes are cumulated over a number of years, which saves a considerable amount of time but means, of course, that they cannot be available until after the end of the period covered. Some prominent examples are *Analytical Chemistry*, vols. 1–30 (1929–58), and subsequently every five years; *The Analyst*, vols. 1–20 (1877–96) and each successive ten volumes; the *Journal of the Chemical Society*, 1848–72 and for successive periods of ten years; the *Journal of the American Oil Chemists Society*, 1917–52; *Chemical Reviews*, vols. 1–60 (1924–60); *Annual Reports on the Progress of Chemistry*, vols. 1–46 (1904–49); the *Journal of Chemical Education*, vols. 1–25 (1924–49) and every ten years; and the *Journal of Polymer Science*, vols. 1–62 (1946–62).

In addition to indexes to individual journals, there are specialist publications which index or abstract a wide range of periodicals, such as *Chemical Abstracts* and others described in Chapter 5.

SURVEYS OF USE

In recent years scientists and librarians have been faced with the problem of dealing with an ever-increasing number of periodicals. Scientists have been confronted with the need to consult a continually extending list of titles as specialization increased and had the effect of scattering the core of papers of real interest through a wider range of journals. Thus in 1975 *Chemical Abstracts* monitored nearly 12 000 titles to abstract its 450 000 documents. Librarians also in times of financial stringency are faced with the twin problems of an increasing number of titles and their ever-increasing cost due to problems of inflation, foreign exchange rates, etc. Much research has therefore been undertaken to determine the basic collection of journals to which each institution would regard it as essential to subscribe and so isolate the 'marginal' journals to be borrowed from the BLLD or other libraries.

In 1975 *Chemical Abstracts* found that 50% of the journal articles it abstracted in that year were provided by 325 journals, 75% by 1384 and 90% by 3589 journals. It is true, of course, that these figures are partly dependent on imponderable factors such as availability and are not in themselves conclusive evidence of the prime importance of these journals.

Interesting figures were also obtained from a survey of requests received at the BLLD in the first three months of 1975 (Bower, 1976). These showed that in spite of a stock of 100 000 titles and a current intake of 45 000, 50% of requests received in all subjects were for 1300 titles and that 34% of these satisfied 80% of the demand. In science and technology 50% of the applications were supplied by 8% of the requested titles, half of them being from the last 3½ years and 81% of those in chemistry being in English. It is difficult to know how to interpret these figures, as we are not told the type of institution borrowing in each case (it can be assumed that industrial libraries do not have the same stocks as universities) or the reason for each application (journals in stock may be away at binding, missing temporarily, etc.).

A comparison was made by Scales (1976) of titles lists produced by citation counting and from use data, the former being the 1000 most cited journals in the last quarter of 1969 listed in the *Journal Citation Reports* (Institute for Scientific Information), and the latter derived from a survey carried out at the NLL in 1969 of requests for issues published in the preceding three years. This showed that *Science* was the only journal to appear in the first five of both lists and that 250 titles were needed before 50% of both appeared. Correlation was generally low, as shown by the following titles common to the first 50 on both lists.

	Citation rank	Use rank
Journal of the Chemical Society	3	16
Nature	4	14
Science	5	2
Journal of Chemical Physics	7	30
Biochimica et Biophysica Acta	8	10

To a large extent these figures are not surprising, as it is to be expected that the most heavily cited journals would be those which libraries already possessed and would be borrowed in the main only by those industrial libraries whose holdings were not very large.

The ISI *Journal Citation Reports* (Anon., 1972a) provide periodically updated rankings of the 1000 most heavily cited journals in a specific period, indicating the number of citations each received, the names of the journals citing it and the dates by year of the articles cited. These statistics are taken from the data base provided by the *Science Citation Index* and to some extent permit the identification of titles required for a general or special core collection, allow for cost/benefit analysis of subscriptions and determine the most important journals of marginal interest to researchers in a special field. There are, however, many limitations to the free acceptance of citations as the only criterion of a journal's importance. Many journals, such as *New Scientist*, are heavily used because of their intrinsic interest but are infrequently cited, and many citations are made more for prestige reasons than for relevance because of the author or subject matter. Familiarity with a given article will also depend to some extent on the circulation of the journal, availability of reprints, inclusion in abstract journals, whether in a foreign language, etc. Nevertheless a number of very interesting articles and lists have been published by Eugene Garfield. Based on a million citations in the *Science Citation Index* for the last quarter of 1969, a list of 50 of the most important journals in all sciences was compiled (Garfield, 1972a) and compared with a list ranked by impact factor – i.e. number of citations divided by number of published items (Anon., 1972b) – which reduces the bias normally favouring journals with large circulations. The first ten in each were:

Most cited	Highest impact factor (Corrected *C.C.* 7 Feb. 1973, No. 6)
1. *Journal of the American Chemical Society*	1. *Advances in Protein Research*
2. *Physical Reviews*	2. *Pharmacological Reviews*
3. *Journal of Biological Chemistry*	3. *Bacteriological Reviews*
4. *Nature*	4. *Annual Review of Biochemistry*
5. *Journal of the Chemical Society*	5. *Physiological Reviews*
6. *Journal of Chemical Physics*	6. *Accounts of Chemical Research*
7. *Science*	7. *Solid State Physics*
8. *Biochimica et Biophysica Acta*	8. *Advances in Enzymology*
9. *Proceedings of the National Academy of Sciences of the USA*	9. *International Review of Cytology*
10. *Biochemical Journal*	10. *Journal of Molecular Biology*

This comparison shows only 11 titles common to both lists and the distorting effect of the many review journals which figure so much in the high-impact list.

Another comparison made from the same data base is between the core literatures of biochemistry and chemistry (Garfield, 1972b) based on the journals most heavily cited by the *JACS* and *Biochemistry*, both ACS publications. A similar comparison made between the *JACS* and the *Journal of Chemical Physics* (Garfield, 1972c) provides lists of the 50 journals most cited by each, the first ten being:

Journals cited by *JACS*	Journals cited by *JCP*
1. *JACS*	1. *JCP*
2. *JCS*	2. *Physical Review*
3. *JCP*	3. *JACS*
4. *Journal of Organic Chemistry*	4. *Journal of Physical Chemistry*
5. *Tetrahedron Letters*	5. *Proc. Royal Society*
6. *Inorganic Chemistry*	6. *Trans. Faraday Society*
7. *Journal of Physical Chemistry*	7. *Molecular Physics*
8. *Chemische Berichte*	8. *Physical Review Letters*
9. *Canadian Journal of Chemistry*	9. *JCS*
10. *Angewandte Chemie*	10. *Acta Crystallographica*

In a comprehensive article first published in *Science* (Garfield, 1972d) and reprinted in *Current Contents* (Garfield, 1973) Garfield concludes that 50% of all references are cited by 152 journals and that 500 journals, mainly interdisciplinary, publish approximately 70% of all articles cited, with the greatest number of citations occurring in the two years subsequent to publication. He lists the 152 most frequently cited journals in rank order, the first ten being:

1. *JACS*	6. *JCP*
2. *Phys. Rev.*	7. *Science*
3. *J. Biol. Chem.*	8. *Biochim. Biophys. Acta*
4. *Nature*	9. *Proc. Nat. Acad. Sci. USA*
5. *J. Chem. Soc.*	10. *Biochem. J.*

The first ten chemical journals on the list with their ranking are as follows:

1. *JACS*	10. *Biochem. J.*
3. *J. Biol. Chem.*	15. *J. Org. Chem.*
5. *JCS*	23. *JPC*
6. *JCP*	24. *Chem. Ber.*
8. *Biochim. Biophys. Acta*	30. *Analyt. Chem.*

In spite of the limitations already mentioned, these lists do provide some objective data for librarians trying to ensure that they are getting the best value for their limited resources. Another approach was chosen at the University of Surrey, where various factors were considered (Panton and Reuben, 1971) to enable a list of chemistry journals to be drawn up showing the ten 'best buys'. These were, in order of cost/ citation ranking:

1. *JACS*	6. *Analyt. Chem.*
2. *Trans. Faraday Soc.*	7. *Org. Synth.*
3. *JCS*	8. *Chem. Ber.*
4. *Chem. & Ind.*	9. *JCP*
5. *Inorg. Chem.*	10. *Chem. Commun.*

GUIDES TO PERIODICALS

The purpose of an abstracting or indexing journal is to help discover the existence of specific articles on a given subject. The next problem, having found them, is to locate the periodicals in which they appear. Because of the difficulty experienced in identifying and procuring scientific journals, various guides have been provided which enable one to check titles and publication data, and to discover in which libraries they are available.

An invaluable subject guide to the world's periodicals is *Ulrich's International Periodicals Directory*, which gives full details of nearly 60 000 periodicals, including title and translation, language, date of first publication, full particulars of price, publisher and editor, sponsoring organization, ISSN (International Standard Serial Number) and Dewey Decimal Classification, special features such as cumulations and indexes, and abstracting and indexing journals which scan each title. It also includes an index to new periodicals and to the publications of international organizations. A companion volume, *Irregular Serials and Annuals*, is an international directory of serials, annuals, continuations, conference proceedings and similar publications issued irregularly or annually, arranged by subject. Full details are given for each entry, including Dewey Decimal Classification and ISSN.

Two other publications which are useful in tracing periodicals are M.J. Fowler's *Guide to Scientific Periodicals* (Library Association, 1966), which is an annotated bibliography of over 1000 guides to periodicals, and D. Woodworth's *Guide to Current British Journals* (Library Association, 2nd edn, 1973). The first volume is a classified list with full bibliographical data and includes intellectual level and coverage; three appendices list journals carrying abstracts, discontinued journals, and societies and their publications. A second complementary volume contains an alphabetical listing of publishers with their addresses and the titles they issue. An extensive list of scientific, technical and professional journals from the developing countries of the Commonwealth is provided by the *Commonwealth Directory of Periodicals*, published by the Commonwealth Secretariat in 1973.

The most important tool for chemists, which is invaluable for its bibliographical information and library locations, is the *Chemical Abstracts Service Source Index (CASSI)*, which enables one to identify and locate journals, patents, technical reports, monographs and conference proceedings cited by *Chemical Abstracts* from its inception in 1907 to 1974 and held by 326 libraries in the US and 73 libraries in 27 other countries. Journal references quoted by *Chemisches Zentralblatt* (1830–1940) and *Beilstein* (to 1965) are included. Information contained in over 53 000 entries catalogued by standard library procedures is very comprehensive and is supplemented by a list of the 1000 most cited journals in *Chemical Abstracts*. It is updated by quarterly and annual supplements.

Another useful publication issued by the Chemical Abstracts Service in 1974, in conjunction with the Biosciences Information Service of *Biological Abstracts* (*BIOSIS*) and the *Engineering Index*, is the *Bibliographic Guide* for editors and authors, which lists journal titles as found on the title page, title abbreviation according to the ISO system and ASTM Coden, for 18 500 current and 9200 discontinued scientific and technical journals, with an indication of which of the three services monitors each current periodical. The *Engineering Index* has also published a list of current periodicals abstracted and indexed by *EI*. *Publications Indexed for Engineering* (*PIE*) contains 2147 titles of journals, special publications, symposia, etc., representing 1693 English-language and 454 foreign-language titles from 48 countries.

Japanese periodicals can be traced through the *Directory of Japanese Scientific Periodicals*, published in 1974 by the Library of the National Diet, Tokyo, which contains 7087 titles in classified order, giving the original, transliterated and translated version, the language used and the language of any summaries provided.

A most useful British publication is the *World List of Scientific Periodicals*, which lists alphabetically by title about 60 000 scientific and technical journals published between 1900 and 1960 with locations in approximately 300 British libraries. Besides its value in locating periodicals, the *World List* helps to standardize abbreviations of periodical titles by giving recommended forms which are required in citations by an increasing number of journals. Unfortunately these are not always the same as the ones recommended by IUPAC, who have specified those used by *Chemical Abstracts* listed in *CASSI* and current issues of *Chemical Titles*. Transactions of international congresses with dates and locations are recorded in an appendix.

A similar function is served by the *British Union Catalogue of Periodicals* and supplements, which record over 149 000 periodicals in all subjects from the seventeenth century, giving title changes, volumes and dates and recording their availability for reference, loan or photocopying in 441 libraries. '*BUCOP*' has now incorporated the *World List*, which it continues through quarterly and annual continuations, including a separate annual list of scientific periodicals under their published titles. Libraries are indicated by three-letter symbols, of which the first two indicate location and the third the general type of library, followed by a number for various libraries of the same type in the same place – e.g. DB/1 = Royal Irish Academy; LB/U-1 = Loughborough University of Technology. There is a useful appendix of recurring international conferences.

Published lists of periodical holdings of major libraries can be used for checking titles, dates and volume numbers of lesser-known journals, and also for obtaining locations for reference or borrowing purposes. An important example of this type is the Science Reference Library's *Periodical Publications in the National Reference Library of Science and Invention*, which lists over 10 000 titles in the Bayswater and Holborn branches.

Based on these extensive holdings, the Science Reference Library's *Periodicals News*, published monthly, lists changes of title, amalgamations, deceased periodicals and other changes in addition to a selection of new titles added. New periodicals can also be traced through the subject indexes to *Chemical Abstracts* under the heading 'Periodicals' and in the *British National Bibliography*.

A useful guide to ascertain correct titles of periodicals and to show their holdings is the British Library's catalogue of *Current Serials Received*, which contains titles of about 45 000 serials received at Boston Spa. There are two supplements, which list transliterated titles originally published in the Cyrillic alphabet and journals published in cover-to-cover translations.

An important American publication is the *Union List of Serials in Libraries of the U.S. and Canada*, which provides data on over 150 000 journals. This is supplemented by the Bowker publication *New Serial Titles 1950–1970*, which gives bibliographical data and locations of some 260 000 serials which commenced publication after the end of 1949 and are taken by any of over 800 libraries in the US and Canada. Country codes, ISSN and Dewey classification numbers are provided. A further volume cumulates 1971–74. This is updated by monthly, quarterly and annual cumulations. A classified subject arrangement is also available.

Another source in checking (incomplete or erroneous) references to periodicals is the synchronistic table. This gives the volume number corresponding to a given year for a particular periodical. Such tables, covering a wide period and dealing with many important journals, can be found in vol. I of the main work (HW) and supplements (EI) and (EII) of Beilstein's *Handbuch der organischen Chemie*, Dyson's *Short Guide to the Chemical Literature* and Friend's *Textbook of Inorganic Chemistry* (vols. X and VII (1)).

Table 3.1   REPORT SERIES AND INDEXES TO THEM

Reports and indexes shown below are the most significant ones in the field of may not be restricted for security or commercial reasons, that are indexed only should be noted that most report series mentioned are currently available on reports of US origin.

The BLLD only collects open unlimited report literature. Access to security the UK should be made to the Ministry in charge of the particular research, or for Centre (DRIC).

Report series Abstracted in	AD	PB NTIS/PS COM	EPA	NASA (NACA)	British reports	Atomic energy ——— ERDA
Government Reports Announcements & Index (GRA)	All DDC released	All	Selected	NASA origin	Some	USAEC origin
NSA (to June 1976)	Topical*	Topical*	Selected	Topical*	Topical*	All
STAR	Topical*	Topical	Topical*	All	Some	Topical
WGA	Weekly Government Announcements.				Current awareness	
EPA Bibliography	Topical	Topical	All			Topical
IAA	Few	Few	Few	Few	Few	
BLLD Announcement Bulletin	British origin	British origin			All unlimited	
R & D Abstracts	Selected	Selected	Selected	Selected	Within subject scope	
Selected RAND Abstracts	Source RAND		Source RAND	Source RAND		
ZAED	Few	Few	Few	Few		Some
INIS Atomindex	Few	Few	Some	Few		Most
RIE	Few	Few	Few	Few	Few	

*Listing of AD, PB and other reports in *STAR* and *NSA* does not imply public and released by the National Technical Information Service (NTIS). Availability

Table 3.1 Continued

report literature. Of course there are many more series of reports which may or
by the issuing organization, or may be referred to only by contract numbers. It
microfiche, and in many cases only on microfiche. This applies particularly to

classified documents is granted only on a 'need to know' basis, and requests in
restricted reports of US and foreign origin to the Defence Research Information

Atomic energy ——— Other	RAND	Universities, institutes, industry			Remarks
		Major series	*Contract number reports*	*ERIC (ED)*	
Some	Most	Some	Many	Few	Items without availability details may not be available to the public. BLLD holds journal, etc., alternatives of most such items. Most US Government-sponsored translations included
All	Topical	Some	Some	Few	Topical journal/book references also listed
Topical	Topical	Topical	Topical	Few	Topical journal/book references also listed
			SDI selection of reports listed in *GRAI, NSA* and *STAR*		
Some	Some	Some	Some		
		Many	Some		Lists mainly conference papers and articles
British		British	British	British	British reports, translations and doctoral theses in a wide range of subjects, including social sciences and humanities
Selected British		Few	Some		The P and T numbers are TRC identification numbers. Primary report numbers, etc., must be quoted when requesting item from BLLD
	All			Some	Various report series, but RAND P series are mainly journal and conference preprints
Some		Some	Some		Emphasis on conference papers in nuclear field. Now only abstracts available
Most		Topical	Topical		Now offering full subject coverage
	Few	Topical	Topical	All	Educational topics in the widest sense

availability. In general, these reports are not available until listed in *GRA*
of non-NASA documents listed in *STAR* may also be subject to restrictions.

Explanatory Notes to Table 3.1

*Government
Reports
Announcements*
(now combined
with
*Government
Reports
Index*)

A fortnightly publication, which supersedes *US Government
Research and Development Reports (USGRDR)*. Lists Defense
Documentation Center (DDC) documents (AD reports),
Publication Board (PB) reports, Dept. of Commerce (COM)
reports, and other reports released by US government agencies
and their contractors, many foreign reports, also US government-sponsored translations and theses. Until 1966 *USGRDR*
was published in two sections: Section 1 was also known as
Technical Abstracts Bulletin (TAB). Individual issues and
earlier cumulative volumes had subject, author and report
number indexes. In 1966 these indexes were discontinued, and
for 1966–70 *USGRDR* and its predecessor, *Government Wide
Index (GWI)*, were the only indexes to *USGRDR*. When in
1971 *USGRDR* changed to *GRA*, *USGRDR*-Index was renamed *Government Reports Index*. In 1967 a Report Locater
Index was re-introduced into *USGRDR*, and this continues in
GRA. Magnetic tape access available through RECON at SRL
and TRC. (See also page 86.)

*Nuclear
Science
Abstracts*
(now continued
by
Atomindex)

Issued semi-monthly. Each issue had subject, corporate and
personal author, and report number indexes, the last giving
cross-references. Entries in subject groups were in accession
number order. Regular cumulations, including five-year cumulations with very comprehensive indexes and a permanent report number index for 1947–61, 1962–66 and 1967–73.
Coverage of not only USAEC, but also all other national and
international series of atomic energy reports, books and theses,
also translations from Russian, Japanese and European languages, as well as bibliographies and literature searches in many
fields. Subject coverage extends well beyond nuclear energy,
i.e. medicine, engineering, agriculture, etc. BLLD holds
virtually all reports listed in *NSA*. Magnetic tape access is also
available.

*Scientific and
Technical
Aerospace
Reports*

Published semi-monthly. Coverage: space, flight and missile
development, also many fringe subjects, i.e. biology, space
medicine, etc. Coverage of not only NASA, but also all other
national and international series of reports, translations,
theses, etc., of interest to space research. Individual issues and
cumulative volumes have subject, author, contract and report
number indexes. *STAR* lists now many items available only
from the BLLD. Recent innovations include a more detailed
subject breakdown, also on-going research projects. Magnetic
tape access through ESRO/ELDO RECON terminals at British
Library SRL and at TRC. (See also page 86.)

*Government
Reports
Index*
(now combined
with *Government
Reports
Announcement*)

Fortnightly publication, issued parallel with *GRA* (formerly
known as *USGRDR-Index*). Computer print-out of index
sections of *GRA*, also USAEC and NASA items in *NSA* and
STAR. *GRA* items are listed in corresponding issue of *GRI*;
NSA and *STAR* items appear usually 1–2 issues in arrears.
Subject, personal author, corporate source, contract, report
number and accession number indexes. Annual cumulations

only for last two years. To cover the years 1960 to 1969, a comprehensive multi-access index of 44 reels of 16 mm microfilm is available under the title *Defense R & D of the 1960s*. Magnetic tape access available.

EPA Reports Bibliography   Cumulative index of US Environmental Protection Agency reports. Up to April 1973 – PB 223693. Supplement to December 1973 – PB 234215.

WGA   *Weekly Government Announcements*. Current awareness SDI in many subject areas. Issued frequently.

International Aerospace Abstracts   Published semi-monthly in parallel with *STAR*. Similar layout and contents, but with emphasis on conference papers, journal articles, reprints and books in the aerospace field. Both *IAA* and *STAR* have world-wide coverage. Magnetic tape access available through ESRO/ELDO RECON terminals at BL SRL and at TRC. (See also page 86.)

BLLD Announcement Bulletin   Supersedes from 1971 *British Research and Development Reports* (1966–70). *BR&DR* contained British reports from all sources in all subject fields. Renamed in 1971 when coverage extended. The *BLLD Announcement Bulletin* lists not only British reports, but also translations received from UK sources and doctoral theses from 25 British universities. Entries in the *Bulletin* are arranged in broad COSATI group order.

R & D Abstracts   Published semi-monthly. Lists selected current and older reports of British and US origin received at DoI TRC. Entries are arranged in COSATI subject group order. Cumulative indexes (author, corporate source, report number, etc.) are available. These also act as indexes to most reports listed in *BLLD Announcement Bulletin*.

Selected RAND Abstracts   Two cumulative volumes: 1946–62 and 1963–72 + quarterly issues cumulating annually. Wide variety of subjects covered on work performed by the RAND Corporation usually on government grants, on mathematics, social and political sciences, environment, etc. Topical reports also abstracted in other indexes described above.

ZAED Information   Monthly. Lists conference papers presented in the nuclear field. Microfiche service discontinued in mid-1974.

INIS Atomindex   Semi-monthly. Abstracts of reports, conference papers, etc., available from the International Nuclear Information Service of IAEA.

Research In Education   Monthly. Abstracts of Educational Resources Information Center (ERIC) reports originating from 19 Clearinghouses of the US Dept of Health, Education and Welfare. Most documents listed are available on microfiche from the BLLD.

REPORT LITERATURE

Since the war the amount of primary information has increased faster than the ability of the scientific journals to publish it. Furthermore, much scientific research during and since the war has been carried out under government contract, which often requires frequent and detailed progress reports. The fact that much of this information is often both confidential and highly specialized has led to the practice of publishing separate reports, which now rival journals as the main source of primary information. Many of these reports are of American origin, being produced by a variety of sources, from government agencies to private firms. They are usually designated by special codes of numbers and initials, and are often not listed in the standard bibliographies, which makes them very difficult to locate. *Table 3.1*, reproduced by kind permission of British Library Lending Division, shows the series at present available.

Many of the early post-war reports consisted of American wartime research results released for publication by the Publications Board (PB reports). These PB and AD reports were published by the Clearinghouse for Federal Scientific and Technical Information (CSFTI), now the National Technical Information Service (NTIS), which also issues a number of selected bibliographies. The Energy Research and Development Administration (ERDA), formerly the American Atomic Energy Commission (AEC), issues a vast number of reports on all subjects which are also distributed by NTIS. Other important American reports are published by NASA (National Aeronautics and Space Administration), which is also responsible for the important abstracts journal *STAR* (*Scientific, Technical and Aerospace Reports*). This provides a comprehensive survey of report literature. The Rand Corporation of Santa Monica, California, also produces reports (*Rand Reports*) on a wide range of subjects.

Many important reports originate from similar government and industrial sources in Britain. Some of the earliest were the *BIOS, CIOS, JIOS* and *FIAT Reports*, which contained very important information based on enemy research data captured at the end of the war and edited and published by the Technical Intelligence Documents Unit (TIDU). An index in six parts plus various supplements to these reports was completed at the end of the war. Modern sources include the Aeronautical Research Council (ARC), the UK Atomic Energy Authority (UKAEA), the National Engineering Laboratory (NEL), and the British Iron and Steel Research Association (BISRA), which issues an open report list every six months.

Information on reports available sometimes appears in the scientific and technical press, in the abstract journals, in the official list of government publications, or in the lists of reports issued by the Research Associations themselves. Owing, however, to their complexity and the difficulty experienced in identifying and locating them, a number of special guides to report literature has been issued.

The *United States Government Research and Development Reports* (*USGRDR*) was a continuation of the *Bibliography of Scientific and Industrial Reports* (1946—49) and its successor, *Bibliography of Technical Reports* (1949—54). This has now been superseded by *Government Reports Announcements & Index*. Announcements of reports are entered by accession number under 22 subject groups and consist of brief abstracts with descriptors. Indexes provide approaches by subject, personal author, contract grant number and accession/report number. It is published bi-weekly by NTIS, which is a subsidiary of the US Department of Commerce and is the central source in the US for the public sale of government-sponsored research, development and engineering reports prepared by Federal agencies or their contractors. Its present collection exceeds 800 000 titles. NTIS also runs on-line searches of over 400 000 reports added since 1964, and 180 000 projects compiled by the Smithsonian Science Information Exchange. Reports announced are priced and sold either on paper or as microfiches. Current summaries of new research reports are published in *Weekly Government Abstracts*, a series of newsletters, indexed annually. *SRIM* (*Selected Research in Microfiche*) is an automatic bi-weekly service which enables subscribers to receive microfiches of complete research reports in any area selected.

USAEC and UKAEA reports were fully indexed in *Nuclear Science Abstracts* (*NSA*), which had a report number index in each issue and gave details of availability and price. In June 1976 *NSA* was discontinued, as it was felt that with growing international co-operation it was no longer necessary for the United States to bear the entire load of covering the atomic energy literature of the world. The Energy Research and Development Administration (ERDA), which has replaced the USAEC, will continue to provide coverage of the US literature to the International Nuclear Information System (INIS), which will cover the same ground as *NSA* in the *Atomindex*, published by the International Atomic Energy Agency. In place of *NSA* ERDA is publishing a new journal, *ERDA Energy Research Abstracts* (*ERA*), which will contain abstracts and indexes to all literature produced by ERDA and foreign countries on non-atomic energy topics.

British and foreign scientific reports are also recorded in *R & D Abstracts*, published by the Technology Reports Centre (TRC), St. Mary Cray, Orpington, Kent BR5 3RT, which acts as a clearing house for the thousands of reports produced by the Department of Industry and overseas sources. Reports originating from the TRC are available on loan from the BLLD but on sale only from TRC. The TRC also publishes a subject index to NTIS microfiches prepared from the tape version of *Government Reports Announcement & Index*.

British atomic energy reports are recorded in the UKAEA *List of Publications Available to the Public*. The BLLD *Announcement Bulletin* records all British reports received. Most of these reports are available from the BLLD, which has the most comprehensive collection of report literature in the world, including NTIS, NASA, ERDA, RAND, ERIC and many other reports series. The BLLD is a depository in the UK for EEC documents.

CONFERENCE PROCEEDINGS

See Chapter 7.

THESES

Another important source of original information is provided by the reports submitted by candidates for higher degrees at universities on research work carried out under supervision. These theses or dissertations very often appear ultimately in full or in part in conventional scientific journals or in book form and are then recorded as part of the standard literature. According to Bottle (1973), about one-third of a sample of papers from several chemical journals was derived from predated theses. Until this happens they are usually available for reference purposes in the library of the institution in which the work was carried out, or in other libraries by way of inter-library loan. One difficulty is to learn of their existence at this stage. Some universities, such as London, Leeds, Glasgow, Southampton, Exeter, Sheffield, Swansea, Oxford and Cambridge, issue lists of their own theses or record them in their calendars. An indication of others pending can sometimes be obtained from publications such as the ACS *Directory of Graduate Research* or *Scientific Research in British Universities and Colleges*. (See Chapter 16.)

Since 1950 the main guide in Britain has been the Aslib *Index to Theses Accepted for Higher Degrees in the Universities of Great Britain and Ireland*, which records new theses annually by subject with the name of the author, university, title and degree. Proposals have recently been made for speeding up notification of the availability of British theses by producing an indexing journal on a more frequent basis, backed up by microfiche versions of abstracts of the theses indexed.

Doctoral theses of about 360 American and Canadian institutions and a few European universities are arranged under broad subject headings in *Dissertation Abstracts International*, which commenced in

1938 as *Microfilm Abstracts* and is now published monthly in two parts: A (Humanities) and B (Sciences and Engineering). Each issue has a keyword title and author index to its own contents but the author index published annually cumulates both parts. Each part is divided into five sections, those in Part B being IB Biological Sciences, IIB Earth Sciences, IIIB Health and Environmental Sciences, IVB Physical Sciences (Pure and Applied) and VB Psychology. All theses abstracted can be purchased from the publisher, University Microfilms, or the institution concerned. A cumulative work covering 150 000 doctoral dissertations abstracted in Vols 1–29 was published by University Microfilms in 1970. It is arranged in nine volumes: Vol. 2 deals with chemistry and Vol. 9 is the author index. In July 1967 University Microfilms introduced DATRIX (Direct Access to References Information, a Xerox service) system. DATRIX makes a computerized search of University Microfilms' dissertation files which 'contain 95% of all dissertations recently written in U.S. and Canadian universities as well as thousands written since 1938'. University Microfilms also publishes *American Doctoral Dissertations* (listing all doctoral dissertations accepted by US and Canadian universities since 1963) and *Masters Theses*.

Doctoral theses recorded in parts A and B of *Dissertation Abstracts International* are also listed by the BLLD, which has already received over 200 000 of them, in its *Announcement Bulletin*. Doctoral theses from many British universities have been borrowed and microfilmed since 1971 by BLLD, which lists them also in its *Announcement Bulletin* and supplies Xerox copies for loan and 35 mm microfilm or Xerox hard copy for retention. Half of all British doctoral theses are now available for loan from BLLD, to which applications will often be made in preference to the university concerned.

The Library of the University of Liege has published several sequences of its *Répertoire des thèses doctorat européennes*, which abstracts doctoral dissertations received by the library through its exchange service, including Belgian theses and those of many British universities, such as Oxford, Reading and Sussex. It is divided into three sections, covering Humanities, Science and Medicine, with a keyword index to each in French, German, English and other languages. Bibliographies of theses appearing as separate listings are recorded in the *Guide to Theses and Dissertations: An International Annotated Bibliography of Bibliographies*, by M.M. Reynolds (Gale Research, Detroit, 1975).

To trace foreign-language theses it is sometimes necessary to consult national lists published in the countries concerned. French theses have been recorded since 1884 in the *Catalogue des thèses et ecrits académiques* and in Supplement D of *Bibliographie de la France*, and

German theses in the *Jahresverzeichnis der deutschen Hochschul-schriften*. A publication of more direct chemical interest is the *Bibliographie der deutschen Hochschulschriften zur Chemie*, of which a number have been published in Leipzig.

Theses of chemical interest are also recorded in *Chemical Abstracts*, while the January number of *Chemical Engineering Progress* records about 400 Ph.D. theses completed in the previous and preceding years under subject headings relating to chemical engineering, from Absorption to Wetting processes.

REFERENCES

Anon. (1963). *Nature,* 197, 426
Anon. (1972a). *Current Contents,* April 19, No. 16, 5
Anon. (1972b). *Current Contents*, Feb. 28, No. 8, 6
Baker, D.B. (1976). *Chem. Engng News,* May 10, 23
Barr, D. (1972). *Trends in Book Production and Prices*, NCL
Bottle, R.T. (1973). *J. Docum.,* 29, 281
Bower, C.A. (1976). *BLL Review,* 4, 31
Cook, J. (1966). *Advan. Sci.,* 23, 305
Garfield, E. (1972a). *Current Contents,* Jan. 12, No. 2, 5
Garfield, E. (1972b). *Current Contents,* Feb. 2, No. 5, 6
Garfield, E. (1972c). *Current Contents,* Mar. 1, No. 9, 5
Garfield, E. (1972d). *Science,* 178, 471
Garfield, E. (1973). *Current Contents,* Feb. 7, No. 6, 5
Haglind, B.J. and Maizell, R.E. (1965). *J. Chem. Docum.,* 5, 158
Kean, P. and Ronayne, J. (1972). *J. Chem. Docum.,* 12, 218
Panton, D. and Reuben, B.G. (1971). *Chem. in Britain,* 7, 18
Scales, P.A. (1976). *J. Docum.,* 32, 17
Senders, J. (1977). *Information Scientist,* 11, 3
Wootton, C.B. (1976). *BLL Review,* 4, 41

4

Translations and their sources, with special reference to Russian literature

C. R. Burman

If at one time there was a tendency for some scientists to ignore the foreign chemical literature in the belief that any worthwhile research would eventually be published in English, there can be no doubt that today important scientific developments taking place in the Soviet Union and in many other countries formerly regarded as technologically and economically backward are described only in the vernacular.

Recent trends in the growth of the scientific literature based on an analysis of documents scanned for *Chemical Abstracts* (Baker, 1976) have shown that, while the use of English as an international language is on the increase, many scientific papers are still published in foreign languages. As far as countries are concerned, the United States still publishes the largest percentage of papers, and though these have shown a numerical increase from 63 267 in 1970 to 81 697 in 1975, these figures represent a percentage decline from 28% of the total in 1970 to 25.8% in 1975, which is the same as the 1929 figure, after having risen to 30% in 1966. The Soviet Union, whose percentage output was 3.4% in 1929 and 21.3% in 1966, rose to 25.3% in 1974. Output from the Comecon countries as a whole averaged 32%. In 1973 *Chemical Abstracts* even began to index journals received from mainland China after an interval of six years since the Cultural Revolution.

A similar analysis by language showed a steady increase in the use of English as a percentage of the total number of documents scanned, from 43% in 1961 to 59.7% in 1975, the total share of the English-speaking countries being 40%. The main foreign languages, apart from

Russian, are still German, French, Japanese and Italian, though all of these have shown a steady decline. Italian has now been replaced as sixth language by Polish, while close on 60% of the research carried out in Japan is reported in languages other than Japanese. A comparative table (*Table 4.1*) shows the percentages.

Table 4.1

	1961	*1965*	*1970*	*1975*
English	43.3	52	56.4	59.7
Russian	18.4	20	22.6	23.3
German	12.3	9.8	6.6	4.8
French	5.2	5.1	4.0	3.0
Japanese	6.3	4.0	3.4	3.0
Polish	1.9	1.9	1.1	1.2
Italian	2.4	1.9	1.4	0.7
Others	10.2	5.3	4.5	4.3

Although the figures show that an English-speaking chemist can read 60% of the world's scientific publications in his own language, a considerable proportion remains as Russian originals, though a survey carried out at BLLD (Wood, 1966) showed that less than 10% of British scientists claimed a knowledge of the language, and the proportion has certainly not increased since then, as Russian language teaching has suffered a severe decline in British schools in recent years. Evidently, then, a special case can be made out for Russian literature, and special measures have proved necessary to deal with it. These have included more detailed abstracting, comprehensive reviews, formation of translation pools for individual translations, publication of collections of selected articles on a given theme from various Russian journals in English translation and the provision of complete translations of individual journals.

INDIVIDUAL TRANSLATION

For a translation to read fluently and yet be accurate, linguistic knowledge and subject knowledge must be combined. As the latter is harder to acquire than the former, the chemist should be prepared to make an effort to understand the basic grammar and vocabulary of at least one foreign language, to find out which are the best dictionaries and how to use them. Even the ability to transliterate the Cyrillic alphabet is well worth acquiring. Often the use of formulae and structural diagrams combined with ability to translate captions will provide the gist of the article concerned and obviate the need for a full translation. A service of this kind — particularly for German, Japanese, Russian and other Slavonic languages, Spanish, Portuguese and the Scandinavian languages — is provided by the Science Reference Library of the British Library, in which a linguist co-operates with the reader to give an oral translation of parts of an article to enable the latter to decide which if any sections are relevant and merit full translation. A comprehensive guide to multilingual, French, German and Russian dictionaries, with a detailed bibliography of books and articles on scientific translation and foreign-language material, was presented some years ago at a symposium (Anon., 1970). This also included papers on language learning for chemists, possible pitfalls in translation, and some sources of information for Russian and French. Other relevant aids include advice on how to scan chemical papers in Russian and Japanese (Reid, 1970), on finding the meaning of abbreviations (Wohlauer and Gholston, 1966) and on identifying a wide range of foreign languages (Piette, 1965).

If a full translation is required and has to be specially commissioned, care will be necessary in the selection of a translator, as the process is extremely expensive, though of course it may often be cheap in the end if it avoids unwanted duplication of research. Two reputable organizations in the UK maintain current lists of approved translators. One is Aslib, with its *Register of Specialist Translators*, which enables the linguistic competence and subject knowledge of the translator to be readily assessed; and the other is the Institute of Linguists, with its *Index of Members of the Translators' Guild*, which is updated by new editions plus quarterly supplements. The arrangement is a classified one and lists under each subject the names and addresses of translators and their languages. There are also language and geographical listings. The 1973 edition quoted the following approximate scales of fees: French £5, German £5.50, Russian £8.50 and Chinese £11.50 per 1000 words. A useful guide is available for the UK (Millard, 1968). Similar guides to translators in the USA are also available (American Translators Association, 1971; Kaiser, 1965).

One method which avoids the problem of securing a competent freelance translator is the service provided by the British Library Lending Division (BLLD). Originally intended for translations from Russian scientific literature, the scheme now includes the social sciences and humanities, and covers books and periodical articles in most languages except French, German, Italian and Spanish. The principal stipulations ensure that the service covers publications of the last two years only, in the case of science and technology, and that these report original research, that they are not already available in translation, and that the translation is for private use only. About 1200 articles and 5—10 books per year are dealt with under this scheme. The applicant for a translation has to agree to edit the draft version and may have to pay a charge varying with the subject, language and type of printing used for the finished product. In March 1973 as a short-term trial a preliminary stage was inserted before a full translation was undertaken. This involved sending the requester the original article with an English summary from an abstracts journal or other source, if available, or, if not, with a translation of any conclusions or summary section and captions to tables and diagrams. The applicant was then asked whether this information was adequate for his purposes or whether a full translation was necessary and could be justified. Results seemed to justify the continuation of this system.

TRANSLATION SERIES

It is sometimes possible to obtain a translation of a given article other than by commissioning it directly, as there are many organizations responsible for the production of translations which list and make them publicly available. These specialized sources are very varied in range, and include the Atomic Weapons Research Establishment, whose translations from Russian and other languages cover a wider field than might be assumed; the Canadian Defence Research Board, which publishes a series of translations on broad subject fields for each of a large number of languages; the Energy Research and Development Administration, formerly the United States Atomic Energy Commission; the Royal Aircraft Establishment's *Library Translations*; the Building Research Station's *Library Communications*; the Coal Tar Research Association's *Technical Notes*; the *Library Translations* of the Atomic Energy Research Establishment; the *Library Translations* of the Cement and Concrete Association; the translations, mainly from German technical journals, of the British Coke Research Association; the Central Electricity Generating Board; and the translations of the Rubber and Plastics Research Association of Great Britain, which deal largely with technical papers from Eastern Europe.

The Production Engineering Research Association (PERA) provides a translation service for technical, scientific and commercial information from books, journals, reports and patents from and into major languages. A major source of translations connected with the steel industry is provided by the British Iron and Steel Industry Translation Service (BISITS), founded in 1957 by the Iron and Steel Institute and the British Iron and Steel Research Association. Over 750 translations are published annually, with full text, diagrams and bibliographical references, from all the major languages of technical importance, including Chinese and Japanese. BISITS issues weekly lists of translations available with summaries and also publishes annual indexes. Most of the translations are made by member companies, but suitable translations may be undertaken free on request.

Collections of translations from chemical papers have been issued by Consultants Bureau. These consist of articles on specialized subjects culled from Soviet chemical journals translated by Consultants Bureau and presented in symposium form. Volumes in this series have included the following: *Soviet Research on Complexes and Coordination Chemistry*, 1949—56, a selection of 372 articles; *Soviet Research in the Analytical Chemistry of Uranium*, a collection of ten papers from the *Soviet Journal of Analytical Chemistry* and *Doklady of the Academy of Sciences of the USSR* over the period 1949—58; *Soviet Research in Boron Chemistry*; and *Soviet Research in Catalysis*.

TRANSLATED JOURNALS

In the late 1950s, when the demand for Russian scientific literature was at its peak, a large programme was put in hand to provide complete cover-to-cover translations of the most important Soviet journals. It was considered that if only a small proportion of the articles in any particular issue was of interest, it would be more economical to translate the entire contents for publication. This method obviates omissions and facilitates bibliographical control and handling, in that translated issues carry the same number and date as the original, but is very costly and wasteful, as some articles are included which are of no interest to anybody. Furthermore, when a translation is required, the need is usually urgent and the researcher is not normally willing to wait up to two years, which may be the time lag between publication of the original article and its appearance in a cover-to-cover translation.

In the United States cover-to-cover translations of Russian journals are financed in three different ways:

(1) by grants from the government to non-profit-making scientific societies or research organizations;
(2) by direct contracts with translating agencies;
(3) by (unsubsidized) commercial translating agencies (which have to impose high subscriptions to their journals).

The first two methods are largely financed by the National Science Foundation, which also subsidizes a number of books and monographs produced by commercial publishers, and organizes the spending of money, received from foreign governments in return for American food surpluses, on buying translations from abroad. These cover Russian material through the Israel Program for Scientific Translations, and translations from Polish made in Poland, and from Serbo-Croatian and other Eastern European languages carried out in Yugoslavia. These translations, which amount to close on one million pages, are announced by the National Technical Information Service (NTIS) in various publications, such as the *Government Reports Announcements and Index, Weekly Government Abstracts*, the *Fast Announcement Service* and the *Monthly Catalogue of U.S. Government Publications*.

One of the largest commercial translation publishers in the United States is the Plenum Publishing Corporation, which publishes in translation a number of major Soviet chemical journals, including the *Soviet Journal of Bioorganic Chemistry*, the *Soviet Journal of Coordination Chemistry* and the *Soviet Journal of Glass Physics and Chemistry*. Plenum has also undertaken an extensive programme of translation from Chinese covering the medical, biological, chemical, physical and geological fields. Its aim is to provide cover-to-cover translations of the fifteen major scientific journals published by the People's Republic of China since the Cultural Revolution, including *Geochimica* and *Huaxue Tongbao* (*Chemical Bulletin*).

The Ralph McElroy Company, of Austin, Texas, publishes an English edition of *Kobunshi Ronbunshu*, formerly *Kobunshi Kagaku*, under the title *Japanese Polymer Science and Technology*; an international edition in English of the *Melliand Textilberichte*; and the Soviet journal *Khimicheskaya Promyshlennost'* in translation as the *Soviet Chemical Industry*. This firm also offers custom translations from all languages.

In the UK the British Library sponsors ten cover-to-cover translated journals and two journals of selected translations. In conjunction with the Chemical Society it publishes the *Russian Journal of Physical Chemistry*, the *Russian Journal of Inorganic Chemistry* and *Russian Chemical Reviews*. The journal *Coke and Chemistry* is published in association with the British Carbonisation Research Association, and *International Polymer Science and Technology* jointly with RAPRA. This journal, which replaces *Soviet Rubber Technology* and *Soviet Plastics*, consists of abstracts from Soviet, East European and Japanese journals and about one-third of full translations of important articles. Subscribers may propose some of the abstracts for publication in full. At the present time about 270 titles are published in cover-to-cover translations, about 90% of them from the Russian. The Science Reference Library alone subscribes to well over 200 of these, changes in holdings being announced in their four-weekly publication *Periodicals News*. A list of these is given in No. 1 of their Aids to Readers: *Holdings of Translated Journals*, updated at intervals; also in the BLLD catalogue *Current Serials Received*, which provides, in addition, a list of transliterated titles of journals originally published in the Cyrillic alphabet;

and in the annual cumulation of the *World Index of Scientific Translations*. Some of the most important translated chemical journals are listed at the end of the chapter. More comprehensive lists can be found in the guides mentioned.

In addition to the journals which are translated in their entirety, many, including some published in foreign languages, contain abstracts in English either of their own contents or of related journals. *International Chemical Engineering* is a quarterly publication of the American Institute of Chemical Engineers which contains selective translations of current articles; the *Journal of Chemical Engineering of Japan* contains English abstracts of articles in *Kagaku Kogaku Ronbunshu*. An Occasional Publication of the Science Reference Library, *Current Japanese Journals Containing Articles on Pure Chemistry*, lists 163 periodicals, of which 107 contain main articles written in English or sometimes in another European language, and an additional 28 summaries in a European language, even though the contents are in Japanese. Other collected abstracts of foreign literature include *Abstracts of Bulgarian Scientific Literature, Chemistry*; the US Joint Publications Research Service's (JPRS) *East European Scientific Abstracts* and *USSR Scientific Abstracts*, which have now merged into *USSR and East European Scientific Abstracts, Chemistry*; and the Polish Academy of Science's *Quarterly Review of Scientific Publications*. The JPRS, a component of NTIS, has since 1963 issued a series of *JPRS Reports* of translations from Sino-Soviet materials and the more difficult languages, consisting of journal articles or selective translations on a particular theme. Although intended for government agencies, they are also publicly available and are listed in the *Monthly Catalog of U.S. Government Publications* and in the *Catalogue of Current JPRS Publications*. The Derwent Information Service provides a series of abstracts in translation of patents from the USSR, Germany, France, Netherlands, Japan and Belgium.

COLLECTIONS OF TRANSLATIONS

The largest collection of translations in the UK is held by the British Library Lending Division (BLLD), which has been collecting translations from all languages since 1966, prior to which it specialized in Russian, Eastern European and Oriental languages. The collection now totals approximately 300 000 items, to which about 30 000 articles are added annually. BLLD's holdings of US Government-sponsored translations from NTIS, JPRS and the National Translations Center (NTC) at the John Crerar Library in Chicago, with which it has an exchange arrangement, and commercial publishers, universities, industry and government research establishments in the UK are very extensive. A full range of cover-to-cover translated journals also forms part of its stock. BLLD's collection includes the holdings of the old National Lending Library, which specialized in journals, individual articles and monographs in science and technology; and of the National Central Library, which covered social sciences and humanities monographs. Many of these translations are in microform, but all are recorded in the BLLD's card index and are available for loan or consultation. Indexing is under journal title or author in the case of monographs. BLLD receives about 15 000 enquiries annually concerning the availability of translations, and if the item concerned is not held by the BLLD, the enquiry is automatically passed on to Aslib for a check in its *Commonwealth Index of Unpublished Scientific and Technical Translations*, which was set up in 1951 and now contains about half a million entries on card showing the location only, for journal articles, patents and standards, monographs, reports and university dissertations.

A large and comprehensive collection of translations is also maintained by the Science Reference Library. This comprises over 200 cover-to-cover translated journals, mostly from Russian, collections of selected translations, and many series of translations on special topics. Over 38 000 translations from over a score of languages are recorded in the card index at the Holborn Branch, of which 43% are from Russian and 32% from German. Entries are arranged by alphabetical order of the foreign-language title. A record is also made in the volume containing the original article, where this is in the library, to show that a translation is available.

In the United States the National Translations Center (NTC) at the John Crerar Library in Chicago is the chief depository and information centre for unpublished translations in the natural, physical, medical and social sciences from all languages into English. Co-operation is maintained, to avoid duplication, with the European Translations Centre (ETC) in Holland and the BLLD. Items made available by the ETC and BLLD are distributed by the NTC in the Americas, and translations and index cards are interchanged. The NTC will search its files on request, to trace the existence and location of specific translations, and provide photocopies or microforms if available in its collection.

In 1961, under the aegis of the Organization of Economic Cooperation and Development (OECD), the European Translations Centre (ETC) was established in Delft, 70% of the annual costs being met by the Dutch Government and the remainder by the other 16 participating countries. Its function is to provide translations and information about translations of scientific and technical publications from Eastern Europe, the Middle East and the Far East into the languages of Western Europe. The Centre often supplies translations from its own files or can provide details of translators willing to undertake commissions. In addition, translation centres exist in individual European countries which also aim at avoiding unwanted duplication. In the Federal German Republic the Auswertungsstelle für russische Literatur of the Technische Informationsbibliotek, Hanover, collects and publishes translations recorded in its *Neue eingegangene bzw. gemeldete Übersetzungen*, and in the German Democratic Republic the Deutsche Akademie der Wissenschaften published the monthly *Bibliographie deutscher Übersetzungen* from 1953 to 1965, when it went over to a card index. In France the Centre National de la Recherche Scientifique (CNRS) maintains a card index of its translations in addition to publishing the *Catalogue mensuel des traductions*. Its *Bulletin Signalétique* abstracts over 300 Soviet periodicals.

A co-operative effort concerned with translations of nuclear energy literature, known as TRANSATOM, has been established in Brussels by the European Atomic Energy Community, the UK Atomic Energy Authority and the former USAEC, now Energy Research and Development Administration (ERDA). Its main task is directed to translations from the less familiar languages, such as Russian and Japanese, which are listed in the monthly *Transatom Bulletin*. A central card index is also maintained which indicates the availability and price of translations. This index can also be found in many national atomic energy research centres.

GUIDES TO RUSSIAN PUBLICATIONS

The need for a translation presupposes a knowledge of an original work of interest. This knowledge may be derived from references or abstracts, but occasionally it may be necessary to try and trace the original publication of a Russian work, and the following guides may be of help.

Knizhnaya Letopis' (Book Chronicle), the official Russian bibliography, is a weekly classified listing of books just published in the Soviet Union regardless of language, with quarterly name, subject and geographical indexes. It cumulates annually as the indexed *Ezhegodnik Knigi (Book Annual)*.
Novye Knigi (New Books) is a bibliographical periodical providing a weekly classified listing of books intended for publication. Not all are in fact eventually published.
Letopis' Gazetnykh Statei (Chronicle of Newspaper Articles), a weekly index to articles from Soviet newspapers which includes some articles of scientific and technical interest.
Letopis' Zhurnalnykh Statei (Chronicle of Periodical Articles), a weekly publication which indexes articles from Soviet periodicals.

The principal source of secondary publications in the USSR is the Institute of Scientific Information (VINITI), which publishes *Referativnyi Zhurnal*, a series of very comprehensive abstracting journals covering about 1800 Soviet and 12 000 non-Soviet journals.

GUIDES TO TRANSLATIONS

Having found out the titles of articles published in Russian journals, the scientist will next check whether any of them are already available in English. The most important publications for this purpose are:

Bibliography of Translations from Russian Scientific and Technical Literature (Library of Congress, 1953–56)
Translations Monthly (John Crerar Library, 1955–58)
Technical Translations (US Government Printing Office, 1959–67)

Technical Translations served for many years as the central source of information in the United States on Russian and other technical translations available to the public, listing commercial translations obtainable only from agencies as well as monographs translated by government sources. Cumulative author, journal and accession/report number indexes were issued. Each entry indicated the source of origin, the cost, and order number of the translation concerned.

In 1967 the Clearinghouse for Federal Scientific and Technical Information (CFSTI), which in 1970 became NTIS, was made the central source for translations sponsored by the US Government, while the Special Libraries Association (SLA) Translation Center, which shortly afterwards became the NTC, assumed responsibility for translations originating from non-government sources and for the production of a comprehensive index. This is the *Translations Register-Index* (*TR-I*), which is published monthly, with quarterly and annual cumulations. It is in two parts: a Register Section, which gives a full bibliographical description of all new accessions to the NTC in classified order; and an Index Section, which is an alphabetical listing of all journals, reports and patents from which translations have been made. This section includes not only translations recorded in the first part, but also translations indexed by NTIS in *Government Reports Announcements and Index*, those acquired by the BLLD, and others obtained from commercial translators. *TR-I* supplements the *Consolidated Index to Translations into English* (*CITE*), compiled by the NTC and published in 1969. This is a cumulation of 142 000 translations from serially published journals, patents, standards, etc., compiled from *Bibliography of Scientific and Industrial Research* (1946–53), *Bibliography of Translations from Russian Scientific and Technical Literature* (1953–56), the SLA's *Author List of Translations* (1953–54), *Translations Monthly* (1955–58) and *Technical Translations* (1959–67). It consists of two sections: a Serial Citation Index, which is an alphabetical listing of titles according to the system of the *World List of Scientific Periodicals* with translations arranged chronologically under each publication; and a Patent Citation Index arranged by issuing country. Conference proceedings and monographs are not included.

Current translations originating in the UK from government agencies, industry, research associations, universities and learned institutions are recorded in the *BLLD Announcement Bulletin*, a monthly publication which supersedes both *British R & D Reports* and *NLL Translations Bulletin*. Select subject lists of translations are published quarterly in the *BLL Review*, while over a hundred books translated by request and published in offset litho form by the NLL from 1960 to 1972 are listed in the publication *Translated Books* issued in 1973. Books translated from all languages in all subjects are listed in the Unesco annual *Index Translationum*.

The European Translations Centre (ETC) has published a number of useful guides to translations into Western languages from scientific, technical and social science publications in difficult languages. In 1967 publication began of *World Index of Scientific Translations*, now a monthly publication with quarterly and annual cumulations and intended as a finding list of translations of periodical articles and patents, especially from Slavonic languages, Finnish, Hungarian, Chinese and Japanese into Western languages. Its two sections provide an alphabetical listing of items of which translations have been notified to the ETC, and a world index of translations cross-referenced to the listing. In 1971 it absorbed the semi-monthly *List of Translations Notified to the ETC*. The ETC also publishes an annual series, *Translations Journals*, which is a bibliography of periodicals in cover-to-cover translation, abstracted publications and periodicals containing selected translations. It is also a union catalogue showing the availability of translations in Western European libraries, including BLLD and the Science Reference Library.

A very useful publication which provides a comprehensive guide to 278 cover-to-cover translated journals and 53 selective and other journals in translation is the booklet *A Guide to Scientific and Technical Journals in Translation*, compiled by C.J. Himmelsbach and G.E. Brociner, which gives an indication of the interval between original and translation. A useful feature gives Russian abbreviations of Soviet journals mentioned as a key to their full titles.

SOME FOREIGN JOURNALS OF CHEMICAL INTEREST AVAILABLE IN ENGLISH

Angewandte Chemie (International Edition in English). 1962–
Chemische Industrie International. 1956–
Doklady Akademiya Nauk SSSR (*Proceedings of the Academy of Sciences of the USSR*):
   Doklady Chemical Technology. 1956–

Doklady Chemistry. 1956–
Doklady Physical Chemistry. 1957–
Elektrokhimiya (Soviet Electrochemistry). 1965–
Geokhimiya (Geochemistry). 1956–63; selected articles 1964–
Glasnik Srpsko Khemijskog Drustva (Bulletin of the Chemical Society, Belgrade). 1962–
Hua Hsueh Hsueh Pao (Acta Chemica Sinica). 1966–
Itogi Nauki, Seriya Khimiya, Elektrokhimiya (Achievements of Science, Chemistry, Electrochemistry). 1965–
Izvestiya Akademii Nauk Kazakh SSR, Seriya Khimicheskaya (Bulletin of the Academy of Sciences, Kazakh SSR). 1973–
Izvestiya Akademii Nauk SSR, Seriya Khimicheskaya (Bulletin of the Academy of Sciences, USSR. Division of Chemical Science). 1952–
Izvestiya Sibirskogo Otdeleniya Akademii Nauk SSR, Seriya Khimicheskikh Nauk (Siberian Chemistry Journal). 1966–70
Kauchuk i Rezina (Soviet Rubber Technology). 1959–72
Khimicheskaya Promyshlennost' (The Soviet Chemical Industry). 1969–
Khimicheskie Volokna (Fibre Chemistry). 1969–
Khimicheskoe i Neftyanoe Mashinostroenie (Chemical and Petroleum Engineering). 1965–
Khimiko-Farmatsevtickeskii Zhurnal (Pharmaceutical Chemistry Journal). 1967–
Khimiya Geterotsiklicheskikh Soedinenii (Chemistry of Heterocyclic Compounds). 1965–
Khimiya Prirodnykh Soedinenii (Chemistry of Natural Compounds). 1965–
Khimiya i Tekhnologiya Topliv i Masel (Chemistry and Technology of Fuels and Oils). 1965–
Khimiya Vysokikh Energii (High Energy Chemistry). 1967–
Kinetika i Kataliz (Kinetics and Catalysis). 1960–
Koks i Khimiya (Coke and Chemistry, USSR). 1959–
Kolloidnyi Zhurnal (Colloid Journal of the USSR). 1952–
Mekhanika Polimerov (Polymer Mechanics). 1965–
Neftekhimiya (Petroleum Chemistry, USSR: selected articles). 1961–
Ogneupory (Refractories). 1960–
Plasticheskie Massy (Soviet Plastics). 1960–74
Przemsyl Chemiczny (Polish Chemical Industry). 1968–
Radiokhimiya (Soviet Radiochemistry: selected articles). 1959–
Roczniki Chemii (Annals of Chemistry). 1959–60
Teoreticheskaya i Eksperimental'naya Khimiya (Theoretical and Experimental Chemistry). 1965–
Teoreticheskie Osnovy Khimicheskoi Teknologii (Theoretical Foundations of Chemical Engineering). 1967–
Trudy. Institut Elektrokhimii, Sverdlovsk (Electrochemistry of Molten and Solid Electrolytes: selected issues). 1961–
Trudy. Sessiya. Khimiya Sera-Organicheskikh Soedinenii. Soderzhasgchikhysa v Neftyakh i Nefteproduktakh (Chemistry of Organic Sulphur Compounds in Petroleum and Petroleum Products). 1956–
Ukrainskii Khimicheskii Zhurnal (Soviet Progress in Chemistry). 1966–
Uspekhi Khimii (Russian Chemical Reviews). 1960–
Vestnik Moskovskogo Universiteta. Seriya Khimiya (Moscow University Chemistry Bulletin). 1966–
Vysokomolekulyarnye Soedineniya (Polymer Science USSR). 1959–
Zhurnal Analiticheskoi Khimii (Journal of Analytical Chemistry of the USSR). 1952–
Zhurnal Fizicheskoi Khimii (Russian Journal of Physical Chemistry). 1959–

Zhurnal Neorganicheskoi Khimii (*Russian Journal of Inorganic Chemistry*). 1956–
Zhurnal Obshchei Khimii (*Journal of General Chemistry of the USSR*). 1949–
Zhurnal Organicheskoi Khimii (*Journal of Organic Chemistry of the USSR*). 1965–
Zhurnal Prikladnoi Khimii (*Journal of Applied Chemistry of the USSR*). 1950–
Zhurnal Strukturnoi Khimii (*Journal of Structural Chemistry*). 1960–
Zhurnal Vsesoyuznogo Khimicheskogo Obshchestva Imeni D.I. Mendeleeva (*Mendeleev Chemistry Journal*). 1966–

REFERENCES

American Translators Association (1971). *Professional Services Directory*
Anon. (1970). *J. Chem. Docum.*, **10** (2), 106
Baker, D.B. (1976). *Chem. Engng News*, 10 May, 23
Himmelsbach, C.J. and Brociner, G.E. (1972). *A Guide to Scientific and Technical Journals in Translation*, 2nd edn, S.L.A.
Kaiser, F.E. (1965). *Translators and Translations: services and sources in science and technology*, SLA
Millard, P. (1968). *Directory of Technical and Scientific Translators and Services*, Crosby Lockwood
Piette, J.R.F. (1965). *A Guide to Foreign Languages for Science Librarians and Bibliographers*, Aslib
Reid, Emmet E. (1970). *Chemistry through the Language Barrier*, Johns Hopkins Press
Wohlauer, G.E. and Gholston, H.D. (1966). *German Chemical Abbreviations*, SLA
Wood, D.N. (1966). *Chem. in Brit.*, **2**, 346

APPENDIX: SOME USEFUL ADDRESSES

Aslib
3 Belgrave Square, London SW1X 8PL

BISITS
Iron and Steel Institute, 4 Grosvenor Gardens, London SW1

British Library Lending Division
Boston Spa, Wetherby, Yorkshire LS23 7BQ

Institute of Linguists
91, Newington Causeway, London SE1 6BW

Science Reference Library, Holborn Branch
25, Southampton Buildings, Chancery Lane, London WC2A 1AW

European Translations Centre
Doelenstraat 101, Delft, Netherlands

National Translations Center
The John Crerar Library, 35, West 33rd Street, Chicago, Illinois 60616

5

Abstracting and information retrieval services

R. T. Bottle and *B. J. Kostrewski*

In this chapter we are particularly concerned with finding our way in the vast mass of primary material which was described in Chapter 3. It is now quite impossible for any person to read or even skim through the titles of more than a fraction of the newly published periodical literature which may contain something of interest for him, however specialized his field may be. At present *Chemical Abstracts*, to which reference is made in more detail later, selects articles for abstracting from nearly 14 000 current periodicals throughout the world and it is probable there are a great number of lesser or marginal periodicals that are not even covered by *Chemical Abstracts*, particularly from the Iron Curtain countries.

Eventually, much of the information which makes its primary appearance in periodicals, etc., will be culled into handbooks and treatises which are virtually comprehensive at the time of issue, with, from time to time, the publication of new editions covering yet later information. One may regard these sources of information as the distilled and highly fractionated final product from the mass of information that has appeared in the preceding periodical literature.

The provision of an information service on the vast amount of literature currently published and an information retrieval service to cover the gap before the information is finally distilled into such works as *Beilstein* or specialist treatises and monographs is essential if the chemical literature of the past decade or so is to be used as fully as possible by the individual chemist. Traditionally this has been through abstracts, though reviews (Chapter 7) are again increasing in popularity and, particularly where speed and timeliness are very important, title indexes are used.

It is very important to emphasize two different classes of abstracts — namely indicative abstracts, prepared quickly for alerting purposes, and informative abstracts, prepared for a comprehensive abstracting and indexing service such as *Chemical Abstracts*. If thorough and detailed abstracts are required, then it is almost impossible to provide a quick news service, and there are inevitably delays amounting to months or even years.

As no service can provide a complete coverage, it is important to estimate what fraction of the relevant literature may be missed through incomplete coverage. Ninety per cent of the abstracts in *Chemical Abstracts* come from just over 2000 journals (Anon., 1967) and Cahn (1965) found that 50% of the papers on electrochemistry noted by *Current Chemical Papers* in 1963 came from only eleven journals (out of a total of 376 journals scanned). A theoretical explanation for this phenomenon is given by Bradford's well-known Law of Scattering (Brookes, 1977), and a way of estimating the degree of scatter in one's field of interest is given in the section on the *Science Citation Index* (see page 80).

THE COMPOSITION AND LAYOUT OF AN ABSTRACT

An abstract is a summary of the publication or article accompanied by an adequate bibliographical description to enable the publication or article to be traced.

In practice, with some minor deviations, the complete abstract is laid out with the following information.

(1) *The title* in the original language and/or in English. Normally the title of the paper, but uninformative words may be omitted, e.g. improvements in, process for the, studies of the, etc. The title may be slightly expanded if it is not sufficiently informative.

(2) *The names of the authors, including their initials.* Their address is a useful addition if the article is in a rather inaccessible periodical and the reader may wish to send for a reprint. In the case of patent specifications both the inventors and the assignees (normally the employing company) are given.

(3) *The bibliographical reference in the following form:* The name of the periodical abbreviated in a standard form, the year, the volume* and the inclusive pages. In the case of patent specifications, the serial

*If the journal's pagination starts anew with each issue, the issue number (in parentheses) must be given.

number (preceded by an abbreviation for the country of origin, together with the effective date of the patent). In British and foreign patents the date is the date of application, with the addition of a convention date if there is one. For US patents the date is that of the official gazette containing the claim. Date of filing may also be given.

(4) *The body of the abstract.* This may be a reasonably effective substitute for the original if it is a well-written informative abstract; alternatively it may be merely an indicative abstract or synopsis.

(5) Optionally *the name of the abstractor* and *the number of pages, illustrations, etc.* For review articles the number of references is a most useful addition.

Chemical Abstracts Service publishes an authoritative guide, *Directions for Abstractors* (1967).

EARLY ABSTRACTING SERVICES

Long before the advent of *Chemical Abstracts* in 1907, the *Journal of the Chemical Society* and the *Journal of the Society of Chemical Industry* carried abstracts from as early as 1871 and 1882, respectively. In 1926 the abstracts sections of these two journals were combined to form what became *British Abstracts*, which ceased publication in 1953. The history of *Chemisches Zentralblatt* goes back as far as 1830. It was originally published by the Berlin Academy as *Pharmaceutisches Central-Blatt* (1830–49) and was superseded by *Chemisch-pharmaceutisches Central-Blatt* (1850–55); it then became *Chemisches Central-Blatt* (1856–1906) and until 1970, when it ceased publication, was known as *Chemisches Zentralblatt*. It is often stated that the prewar coverage of *CZ* was superior to that of *CA*, and a recent investigation covering chemical kinetics and terpene literature between 1906 and 1938 showed that abstracts in *Chemisches Zentralblatt* were significantly longer than the corresponding abstracts in *Chemical Abstracts* (Atsu, 1976). For that period the coverage by each of the three abstracting services *CZ*, *BA* and *CA* was in the range of 90.3–98.4%, thus leaving the choice of the service to be determined according to language competence, though if length of abstract can be considered to be an indication of information content, *CZ* abstracts would be more informative. These services provide the major key to the earlier literature except where it has been distilled into *Beilstein, Gmelin*, etc.

CHEMICAL ABSTRACTS

Chemical Abstracts is now the only English-language service abstracting all fields of chemistry and chemical technology. Up to 1953 *British*

Abstracts also covered all fields of chemistry but the indexes were never as comprehensive as those of *Chemical Abstracts.*

Chemical Abstracts is designed essentially for the retrieval of information and not as a current news service. The delays in issuing abstracts and subsequently indexes, particularly subject and formula indexes, reduce materially the value of *Chemical Abstracts* for searching for recent information, and a number of supplementary services are being developed to cover this gap.

Chemical Abstracts was started in 1907 by the American Chemical Society (replacing the *Review of American Chemical Research*, 1897–1906, which had a very limited coverage) and, up to 1967, used to appear twice a month. Up to 1960 the (nominally) December issues were devoted to the Annual Indexes: viz. Author, Patent number, Formula and Subject. The first two have always appeared quite punctually and were published annually up to 1962. In 1960 it was decided to publish the Formula and Subject Indexes semi-annually and since 1962 (when annual volumes gave way to half-yearly ones) all the above indexes have appeared at the end of each volume. In 1957 an Index of Ring Systems for each volume was introduced. In 1963 a Corresponding Patent Index or Patent Concordance was introduced, in addition to the Numerical Patent Index, and this gives a reference to the original patent's abstract. A Hetero-Atom-In-Context Index (HAIC Index) was included in Vols 66–74 (1967–71). A Registry Number Index was produced for Vols 71–75 (1969–71). It was replaced by the seven-volume *Registry Handbook–Number Section* (1965–71), for which annual supplements are issued. The Subject Index (1907–72) was split into a Chemical Substance Index and a General Subject Index which includes all headings which do not refer to specific chemical substances, e.g. halogenation, ultracentrifuge, etc., or generic terms.

It cannot be too strongly emphasized that the value of any abstracting service as an information retrieval system depends (apart from completeness of coverage) on the quality and promptness of appearance of the indexes, especially the Subject and Formula Indexes. The completeness of the Subject Indexes was difficult to fault but they have been three years late. They now appear about six months after the end of the volume.

With a view to saving the searcher time, Decennial Indexes have been compiled covering the periods 1907 to 1916, 1917 to 1926, 1927 to 1936, 1937 to 1946, 1947 to 1956. These five Indexes refer to two million abstracts of papers and patents. The 1947–56 Decennial Indexes ran to nineteen volumes and covered 543 064 papers and 104 249 patents. Quinquennial Indexes now replace them. The first, covering 1957 to 1961, ran to a thousand pages more than the last Decennial Index and covered 519 841 abstracts. The seventh Collective

Index (1962–66), covering nearly a million abstracts, ran to 24 volumes and the eighth ran to 35 volumes. The ninth Collective Index (1972–76) covers about 2 million abstracts and runs to over 60 volumes. The tenth Collective Index (1977–81) will run to over 80 volumes and is already being marketed on the instalment plan.

During a five- or ten-year period, standardization of nomenclature in a new or rapidly expanding field develops. Within a five- or ten-year index it will be consistent, though the nomenclature may have been evolving throughout the ten constituent indexes; thus cumulative indexes are somewhat more than a mere time-saving compilation.

In addition to the above cumulative Indexes, two other Collective Indexes have been produced — the Collective Formula Index (1920–46) and the Collective Patent Number Index (1907–36). Since 1936 patent number indexes have been incorporated in the Decennial Indexes.

A list of periodicals abstracted has been published every five years. This alphabetical list or index gives much useful information, such as frequency of appearance, etc. (see page 38 and cf. the *World List* and *BUCOP*). It also gives the (IUPAC) standard abbreviation for each periodical. These abbreviations are obligatory for literature citations in all American Chemical Society journals as well as many others. The latest is called the *Source Index 1907–1974* (*CASSI*) and is available both in printed form and on magnetic tape for libraries wishing to use the list in conjunction with computer-based records systems. Quarterly supplements are issued.

In 1965 a microfilm edition of *Chemical Abstracts* was introduced. This enables one to obtain the six million abstracts produced since 1907 in about 250 cassettes, each not much bigger than a large cigarette packet. One also requires a set of collective indexes (also available in microform) and a reader printer. The latter can be programmed to select the required column automatically (manually selecting models are also quick to use with practice), and after the abstracts have been read, a print can quickly be made by pushing a button. The microfilm version is available on a subscription basis, which includes updating every six to eight weeks with a magazine of microfilm containing 30 000 abstracts and which is somewhat cheaper than the normal subscription. It seems ideally suitable for the newly established library where space is at a premium, provided that some of the alerting services described in a later section are also taken.

Also in 1965 *CAS* commenced a programme which produced a working computer-based information system by 1969. This produces both printed publications and machine-searchable files from a single human processing of primary sources. In this system, titles and abstracts of papers, reports and patents are selected and prepared by chemists, and are then encoded on magnetic tape through a keyboarding process. Computers are then programmed to select and organize this material and to compose pages for printing. Thus a single key-boarding of the information replaces the multiple human handling which used to be necessary for the publication of abstracts, indexes, etc., and permits numerous publication and (magnetic tape) services to be derived from it. Several of these publications and services are alerting services (q.v.) and those currently available are described later.

Some five to six million distinct chemical compounds (about 80% of them organic) are known and information on their nature, properties and uses comprises perhaps 85% of man's chemical knowledge. It is desirable that such data can be manipulated by a computer. To this end the Chemical Compound Registry has been set up. Each compound is assigned a unique 'address', its Registry Number, under which a file containing information on the compound's known names, literature references, etc., is organized. This number is based on a unique and unambiguous machine language representation of the compound's two-dimensional structural formula with a full stereochemical description. The Registry therefore provides a data base which can be machine-searched (at various levels of specificity) for any particular structural feature. All compounds indexed by *Chemical Abstracts* since January 1965 have been registered and by 1978 about four million had been entered. The Registry System is obviously a potentially important alerting and information retrieval tool, and services based on it are available. Provisions have been made for registering a proprietary compound so that its owner may be alerted to related published data passing through the system, but the proprietary data are protected from unauthorized access. Already certain firms are using catalogues which quote the compound's Registry Numbers where available.

Computer-readable files available

CAS (or its agents) provide a number of computer-readable files for incorporation into in-house services. All are available in standard format and are licensed to users. Prospective users can get an evaluation package for each service.

CA Condensates (*CACon*) corresponds to the documents abstracted in an issue of *CA*. Each week it contains full bibliographical details, including names and affiliations of authors, patentees and patent assignees, *CA* Section numbers and cross-references and the Keyword Index entries. Files from mid-1968 are available.

Chemical Titles (*CT*) files are available from 1962, though the standard format only applies to 1972 on. These two-weekly tapes correspond to the information contained in the printed version (pages 82–83).

Selective abstracts files based on groupings of *CA* Sections are marketed. They all contain full abstracts, Keyword Index and Volume Index entries (including Registry Numbers). With dates of earliest available issue they are: *Chemical-Biological Activities* (*CBAC*), 1965; *Polymer Science Technology* (*POST*), 1967; *Ecology and Environment*, 1975; *Materials*, 1975; and *Chemical Industry Notes* (*CIN*), 1974, which contains about 50 000 abstracts per year from 80 key trade periodicals, its weekly tapes corresponding to the printed issues.

Tape versions of *CASSI* (q.v.) and the *Patent Concordance* (from 1962) are also available.

CA Subject Index Alert (*CASIA*), available fortnightly since 1972, contains the Volume Index entries. Thus *CASIA* + *CACon* is in effect all the data in *Chemical Abstracts* less the texts of the abstracts.

UKCIS

Several CAS-affiliated centres have been set up in the US to provide local computerized searching facilities and such facilities are also available in Canada and Sweden. In the UK, UKCIS (United Kingdom Chemical Information Service) was established in 1969 under the auspices of the Chemical Society and based at Nottingham University. UKCIS is now the UK partner of CAS and thus has sole distribution rights for CAS publications and services as well as providing input of abstracts from British chemical journals. (UKCIS also provides a service based on *Biological Abstracts* and *Bioresearch Index*.)

The oldest of UKCIS services is a current awareness service operating on the principle of SDI or selective dissemination of information. Clients' profiles are updated at intervals to reflect their current information needs; output is monthly. In addition, a variety of other services aimed at groups rather than individuals are provided. The popular *Macroprofiles* are being replaced by *CA Selects*, comprising a wide range of broad topics, including Chemical Hazards, Pollution Monitoring, Fungicides, Prostaglandins. Four review services aim to provide a comprehensive service based on review material in *Chemical Abstracts*. These services were developed with the awareness of the important role of reviews in the initiation of research projects and stimulation of creative thinking. *CA Review Index* (*CARI*), a six-monthly KWIC index to all reviews included in each volume of *Chemical Abstracts*, is available with back issues from 1975 (Vol. 82). CAS and UKCIS are running a number of experimental services, such as obtaining the Registry Number for a named chemical structure and vice versa, searches on specific chemical substances and substructure searches back to 1967.

Detailed arrangement of individual issues of *Chemical Abstracts*

For many years up to 1962 each issue of *Chemical Abstracts* was divided by fields of interest into 31 sections, with two of the large sections, organic chemistry and biological chemistry, each subdivided into several sections. The number of sections was then increased and rearranged to bring similar sections together.

There are currently 80 sections and individual subject groupings are available separately as follows: Biochemistry (1—20), Organic Chemistry (21—34), Macromolecular (35—46), Applied Chemistry and Chemical Engineering (47—64) and Physical and Analytical Chemistry (65—80). *CA Biochemistry Sections* was the first to be published separately (1962). These sections are available relatively cheaply and each carries the Keyword Index to the complete issue.

Since 1967 (Vol. 67) the formerly twice-monthly issues have been split into two parts published on alternate Mondays. Sections 1—34 appear one week and Sections 35—80 the next, each with its appropriate Keyword, Patent and Author Index. Each complete issue (Sections 1—80) now runs to about 1200 pages and contains over 9000 abstracts. Since 1967 abstracts have been numbered serially throughout each volume and are still printed in a two-column format. From 1934 to 1967 the columns (not the pages) were numbered and the abstracts unnumbered. Each column was divided into nine sections, labelled (a) to (i) for easy use in tracing abstracts. In indexes and elsewhere an abstract was referred to by its column number and a letter indicating the fraction of the column in which the authors' name or indexed information appeared, e.g. 2149*c* (i.e. one-third of the way down column 2149). Prior to 1947 superscript numbers 1 to 9 were used in place of these letters. From Volume 67 with the individually numbered abstracts, each (abstract) number is followed by a letter (A to Z) which has nothing to do with its position in the column. These suffixes have been assigned by the computer on the basis of the sequence of digits in the number and serve as a check character to prevent errors.

Within each section the abstracts are arranged:

(1) published papers (including reviews) grouped into (unlabelled) sub-classifications;
(2) books, monographs, etc., listed by title, author and publisher, e.g. ASTM standards are entered here;
(3) patent specifications;
(4) a small section giving cross-references by title and (since 1967) abstract number to other sections where there are specific entries having some relevance.

Each weekly part has an author index (in 1962 diacritic marks were abandoned, so ä, ö, ü, ϕ, etc., are now written as ae, oe, ue, oe, etc.), numerical patent index, patent concordance and Keyword index. The latter was introduced in 1963 and lists keywords in both the title and text of the abstract. This and the other weekly indexes are computer-produced. A list of new and discontinued, etc., journals follows the indexes in even-numbered issues.

A new graduate would do well to read the table of contents in any issue of *Chemical Abstracts* and memorize the numbers of the sections of most interest to him.

Methods of using *Chemical Abstracts* and its indexes

There are so many purposes for which the indexes can be used that only a series of suggestions for general guidance can be proffered. The indexes may be used for searching out specific compounds or for making a full literature survey, or for finding equivalent or corresponding patent specifications. They may be used to see whether work on a narrow topic has been done before, whether a new method of analysis has been anticipated, or whether a particular reaction has been described. The following notes are intended to assist the reader to find out how the indexes are constructed, how to use them and how to keep personal notes.

(1) Work backwards when searching the index, i.e. from the later date to the earlier, unless there are very good reasons to the contrary. (If older work is being sought, work backwards from the latest effective date, e.g. earliest preparation of a compound.) The reason for this is that *the latest reference may contain in convenient collective form many or all of the earlier references*. It is particularly important when a literature survey is being undertaken, since a recent review article may fully or partly reduce the task. In certain fields complete bibliographies have been published (see Chapter 7).

(2) Use the most appropriate indexes first, remembering that author, formulae and patent number indexes are far easier to use than the subject indexes. If, therefore, the answer can be got from these indexes, use them first (possible snags in using an author index are discussed on page 33). For example, if chemical compounds are sought, use the formulae indexes which are compiled according to the Hill (1900) system. (NOTE: The Hill system arranges symbols alphabetically, except that in carbon compounds C always comes first, followed immediately by H. Thus organic compounds are arranged C, H, with other element symbols alphabetically and with the lower number of atoms preceding

the higher, i.e. C before C_2, before C_3, etc., then H before H_2, before H_3, etc. The arrangement of the formulae is in the order Carbon + 1 different atom, Carbon + 2 different atoms, Carbon + 3 different atoms, etc., then by number of atoms, i.e. C_1 before C_2, before C_3.) Isomers are listed alphabetically under a given molecular formula from which the desired compound's name may be selected, often more readily than it can be generated before using the Chemical Substance Index or the old Subject Indexes. These should also be checked under the preferred CA Index name, as this often yields many additional references.

(3) Decide on headings under which to search and set these down. Add to this list any headings which may become necessary during the search.

(4) List systematically in columns the references to the abstracts in some form of numerical system. List the indexes to be searched and indicate or cross out each index as searched, then refer to the abstracts themselves and cross out each as seen. If the reference is of value for the particular search, use separate cards (or small sheets of paper) to list the references and the abstract or extract therefrom. These separate cards on the particular search can be arranged later for convenience. Where the original is to be consulted (and this must be done certainly for all important references), any further information can be added on the card. The use of punched cards is strongly recommended on page 87.

(5) Use as much as possible the collective and decennial indexes. This means that up to 1971 can be covered by the collective indexes; from there onwards annual indexes must be used, except where the ninth cumulative index (1972 to 1976) has already been issued.

(6) Before using each of the indexes – chemical substance, general subject, author, formulae, patent number – read the introduction to those indexes. This is particularly necessary in the case of the subject indexes (and these will probably be the type most used for searching for material other than specific compounds). (The subject indexes provide a most useful discussion on chemical nomenclature, especially 1916, 1937 and 1945, while the January–June 1962 one carries a 98 page supplement on this.)

(7) The construction of the subject indexes, in particular, needs understanding. In general, the system is alphabetical; there is a considerable amount of cross-referencing under possible alternative names. Organic compounds are indexed on the basis of parent compounds or index compounds (root names) with substituents following. Thus, 2,5-dichloroaniline was indexed under aniline, 2,5-dichloro, though since 1972 it has been indexed under benzenamine, 2,5-dichloro-. It is therefore necessary to know the root names that are used for indexing purposes, even though some of these may not be systematic names in

themselves. If any difficulty is experienced in finding chemical compounds, the formulae indexes should be consulted to find the name of
the compound before consulting the main alphabetical subject index.
Non-chemists will find that a reasonable knowledge of chemistry and
its terminology is necessary for their efficient use. All should refer to
the *Index Guide* volume of the *9th Collective Index* if the compound
sought cannot be found in recent indexes.

(8) It must be remembered in using the index that *subjects* and *not
words* have been indexed and that the *indexing covers the whole article
and not merely the title*. There are, on average, eleven index entries for
each abstract; all chemical compounds will be included as separate
entries in the indexes. There may be several subject entries. Particular
problems arise in a very new and interdisciplinary field, probably due to
lack of well-recognized keywords (Bottle, 1965).

There should be no difficulty in using the indexes to *Chemical
Abstracts* provided that their construction is understood (read the
Introduction to each of Volume and Collective Indexes), but it is
advisable to have a certain amount of practice in their use.

REFERATIVNYI ZHURNAL, KHIMIYA

This Russian chemical abstracting journal was started in October 1953.
Although it has a narrower coverage than *Chemical Abstracts*, it is a
major source of Russian and East European literature. It is also a quick
and comprehensive source of Chinese literature. During 1978 approximately 120 000 abstracts were published. Each issue has a subject index
and an author index. It provides annually a patent index, a subject
index and an author index, arranged alphabetically according to the
Cyrillic alphabet (any original non-Cyrillic names are transliterated into
the Cyrillic alphabet). There is also an annual formula index consisting
of three parts: (1) a molecular formula index; (2) an index of cyclic
compounds; (3) an index of elements. The elements are listed alphabetically and the compounds are listed below in ascending order of
carbon atoms.

The British Library's Science Reference Library publishes a *Guide to
the Reverativnyi Zhurnal*. This lists the various sections of the *Zhurnal*
and gives guidance on transliteration and use of indexes. Other sections
may also contain items of interest to the chemist, e.g. the biological and
metallurgy sections.

BULLETIN SIGNALÉTIQUE

Published in France by the Centre National de la Recherche Scientifique, it is now produced with the aid of the Pascal computer system.

It consists of ten issues annually and in 1978 Section VII, Chemistry, carried approximately 38 000 abstracts. Each issue has an author index and a substance index. The indicative abstracts are grouped according to the following broad subject categories: General and Physical Chemistry; Analytical Chemistry; Organic Chemistry Theory; Organic Chemistry, Preparation and Properties; Molecular Evolution and Origin of Life. Other sections which could be relevant to the chemist are Biochemistry and Biophysics; Atoms, Molecules, Fluids and Plasmas; and Crystallography.

USSR AND EAST EUROPEAN SCIENTIFIC ABSTRACTS

The chemistry sections published by the US Joint Publications service are published monthly and contain abstracts of papers from scientific and technical journals published in the USSR and East Europe. Up until 1973 this publication was known as *USSR Scientific Abstracts*. Each issue is divided into fourteen sections, which aim to cover the entire chemical field.

RINGDOC

This is an alerting and retrieval service for the pharmaceutical industry now operated by Derwent Publications, London. Originally the service was operated as a co-operative information scheme between several German and Swiss pharmaceutical companies; it was then known as the Dokumentationsring and in 1964 it was taken over by Derwent. The service operates by subscription. Four hundred core journals are scanned. There is approximately a six-week time lag between publication and dissemination by RINGDOC. Weekly abstracts books are provided; also available are thematic booklets covering broad subject areas, while profile booklets concentrate on specific subjects. Retrieval is effected by cumulative author and subject indexes. Scanning of the abstracts is effected by: (1) indexing terms assigned from a thesaurus of approximately 13 000 terms — these provide a system of 'codeless scanning', as it is called by Derwent; (2) free terms — these are also thesaurus-controlled but provide more detail and explanation. In the abstracts the index terms are underlined and the free terms are not. Coding is geared to the needs of the pharmaceutical industry; thus drug trade names, chemical names and diseased states (e.g. respiratory infection) are included for searching.

Searching can also be carried out according to the ring code. This is a code which is special to Derwent and is used for converting chemical/pharmaceutical information into punched card or computer-readable form.

SERVICES PERIPHERAL TO CHEMISTRY

Biological Abstracts

This is published by BIOSIS (Biosciences Information Service) and is a bi-monthly publication which concentrates on biological systems; thus physiological chemistry would be covered. Indeed, biochemical and pharmacological material is being increasingly covered. There are five indexes, which cumulate half-yearly:

(1) The Biosystematic Index, which groups abstract numbers by broad subject category. Thus a paper dealing with carbohydrate metabolism in man would appear as PRIMATE HOMIDAE CARB. MET. These appear in blocks, so that all abstracts dealing with a particular broad topic can be retrieved quickly.

(2) The Cross Index consists of some 650 fixed headings, and headings descriptive of the abstract content are arranged in alphabetical order and the abstracts are then sorted and arranged in columns under appropriate headings such that each terminal digit of the abstract number corresponds to the column number. This arrangement helps in the identification of common numbers when combinations of subjects are being searched.

(3) The BASIC Index is a KWIC index. It consists of the title and 'enrichment terms'. The latter are extra terms added by the editors to expand the information content of the title. However, because of formating constraints, the entire entry is not displayed and, indeed, for the same item the entry can vary in different parts of the alphabetical index.

(4) The Generic Index is an alphabetical index of taxa at the generic level; the taxa and genera are grouped together with a contextual indication.

(5) Author Index.

The bi-monthly issues of *Biological Abstracts* and the monthly *Bioresearch Index* (which concentrates on non-serial publications) are available as computer-readable files. UKCIS construct profiles and operate an SDI service based on these tapes.

Index Medicus and related services

Index Medicus is a monthly publication which is an index and does not have abstracts. The citations, consisting of title, author and source reference, are listed under controlled subject headings. These are taken from *MeSH* (*Medical Subject Headings*), a controlled list of 13 000 preferred terms. Through *Index Medicus* and its precursors the medical literature may be searched back to 1879. *Index Medicus* has been computer-produced since 1963 and the computer tapes are now utilized in *MEDLINE* (the on-line service based on *Index Medicus*) and other spin-off services described on pages 85–86. In hard copy each reference appears three times under the three most important keywords assigned to the document by the indexers. *MEDLINE* has superseded *MEDLARS*, the earlier batch processing system. The depth of indexing available on *MEDLINE* is far greater than in *Index Medicus*. As many indexing terms are assigned as are necessary to express the concepts within the documents, and their identification is as specific as *MeSH* will allow. In addition, the contextual environment of a term can be indicated by the use of qualifiers or sub-headings, e.g. adverse effects can be used with the chemical category to denote reports of adverse effects of chemical compounds. There are 60 sub-headings, and the vocabulary is classified into sixteen categories reflecting various aspects of medicine, e.g. anatomical terms, diseases.

TOXLINE is operated by the US National Library of Medicine, and is a computer-readable data base of bibliographical information on toxicology available as an on-line service. The current file *TOXLINE* covers the period 1971 to date and at present contains some 400 000 references, while *TOXBACK*, covering the period 1965–70, contains some 790 000 references. It is a composite file covering material on toxicology and environmental chemicals and pollutants taken from various sources. The main sources are *Index Medicus*, *Chemical Biological Activities* (produced by CAS), *Toxicity Bibliography* (NLM), *International Pharmaceutical Abstracts*, *Abstracts on Health Effects of Environmental Pollutants*, the BIOSIS data base and *Pesticides Abstracts* of the Environmental Protection Agency. The data base contains full bibliographical citations and may contain abstracts and/or indexing, depending on source. Retrieval is based on 'free-text' searching for words in titles, abstract or indexing term fields. The result of a search may be printed at the terminal. The references may be printed with abstracts or indexing terms, where available, or as a bibliographical citation only.

CHEMLINE is an on-line chemical dictionary which links CAS Registry Numbers to chemical substance names and synonyms. Searching for compounds on *TOXLINE* can be carried out using the Registry Numbers: indeed this is the least ambiguous method, but not all the data bases incorporated into *TOXLINE* carry the Registry Number. Some 77 000 chemical substances and 270 000 synonyms are listed in *CHEMLINE*, which is produced by NLM in collaboration with CAS.

CANCERLINE is a data base of cancer literature produced by the NLM. When a search involves organic chemicals it is advisable to use *CHEMLINE* first.

TECHNOLOGICAL SERVICES

Engineering Index (monthly with annual cumulations, 1884–) is the classified index to some 1400 (mainly English-language) journals which can be useful for tracing literature on applied chemistry (especially before the increased technological coverage by *Chemical Abstracts* from 1967 onwards). *British Technology Index* covers a similar field but its usefulness is restricted, as it indexes only British journals.

Science Abstracts is issued in two parts. *Section A: Physics Abstracts* contains some nuclear and physicochemical information. *Section B: Electrical Engineering Abstracts* is of less direct interest to chemists.

Nuclear Science Abstracts is discussed in Chapter 10.

OTHER ABSTRACTING SERVICES

Over 100 services which cover chemistry and chemical technology or some specific aspect of it were listed in *A Guide to the World's Abstracting and Indexing Services in Science and Technology* (National Federation of Science Abstracting and Indexing Services, Washington, DC, 1963). This now out-of-date guide provided the publisher's name and address and information on price, number of abstracts per year, coverage, etc., and geographical and subject (alphabetical and UDC) indexes. Additions to *Abstracting Services in Science, Technology, Medicine, Agriculture, Social Sciences, Humanities* (FID 372, International Federation for Documentation, 1965) are noted in the *FID News Bulletin*. The Euratom publication *Abstracting and Indexing Journals in the Nuclear and Border-line Fields* (1964) listed 483 such journals.

Two useful check lists are produced and frequently updated by the British Library. They are *Abstracting and Indexing Periodicals in the Science Reference Library*, which lists publications under broad subject headings, e.g. organic chemistry, general and inorganic chemistry, physical chemistry, biochemistry and biophysics; and a *KWIC Index to the English Language Abstracting and Indexing Publications Currently Being Received by BLLD*.

Few of these other abstracting services can be regarded as systems for information retrieval — that is to say, they are not normally provided with very detailed indexes. However, within the fields covered, it is possible to search back, either by looking through the main body of abstracts themselves or in the abbreviated indexes.

Some abstracts journals concentrate on work done within a particular country, e.g. *Abstracts of Bulgarian Scientific Literature, Chemistry*. This national type of publication, like the *Referativnyi Zhurnal*, tends to be produced by countries where the collection and dissemination of information is centralized. In the USA and the UK the pattern is more diffuse, though *USGRDR* carries abstracts of US government-sponsored projects.

Most of the numerous research associations publish abstract services within their limited field of interest. Some professional societies and institutes include abstracts in their periodicals, and certain other officially sponsored or supported bodies, such as the Bureau of Hygiene and Tropical Diseases and the Commonwealth Agricultural Bureaux, put out abstract periodicals. There is also an increasing number of commercially produced specialized abstracting alerting services springing up. Some of these can often be extremely useful in following developments in a relatively narrow technological field. One of the promptest with its abstracts is the monthly *Organometallic Compounds* (R.H. Chandler, 42 Grays Inn Road, London W.C.1, 1962–). Indicative abstracts are normally published within one or two months and an individual metal index is issued for each volume (six issues).

INTERDISCIPLINARY INDEXES

The dream of a comprehensive multi-disciplinary bibliographical index is probably almost as old as librarianship itself, yet the ever-increasing flood of literature of all types and subjects must doom any such project to failure. Even an interdisciplinary index to scientific literature is impossible if it seeks to cover all of the 26 000 scientific journals currently published. Some measure of success (say 50% coverage) can be achieved by selecting the more important journals because of the operation of Bradford's Scattering Law. One index even attempts the impossible task of covering the periodical literature of the sciences and humanities. This is *Internationale Bibliographie der Zeitschriftenliteratur aus allen Gebieten des Wissens*, commonly known as '*Dietrich*' or '*IBZ*' and revered by some librarians out of all proportion to its actual usefulness. It is probably most useful in the non-scientific fields, where indexing and abstracting services are less adequate or long-standing. *IBZ* started in 1911 and was in two divisions, German- and non-German-language periodicals, up to 1965; since then a combined edition covering about 1100 journals has been issued. It is an alphabetical index of keywords from article titles issued in two half-yearly volumes. Its coverage, format and delays of up to two years render it quite useless to the chemist, who should only use it if seeking biographical or historical information.

At first sight *Letopis' Zhurnalnykh Statei* (Chronicle of Periodical Articles, 1926–) is an interdisciplinary index. It is an author index to learned and technical journals published in the USSR, but since these are listed under 30 major headings (usually subdivided), it can in fact be regarded as a number of separate subject guides between the same covers. The average delay time appears to be about three months.

The Royal Society compiled a *Catalogue of Scientific Papers* (four series) covering the whole of the nineteenth century. This is now mainly of historical interest but was once the best-known interdisciplinary (mainly author) index – that is, before the publicity department of the Institute for Scientific Information started promoting the *Science Citation Index*.

Science Citation Index

Science Citation Index (SCI) is produced by the Institute of Scientific Information (ISI), Philadelphia. Both hard copy and computer-readable tapes are available.

SCI is based on the premise that the references cited by an author are in some way relevant to that piece of work and thus give an indication of the conceptual environment in which the work was carried out. This idea, developed by Garfield of the ISI, is not dissimilar to the legal concept of precedent, and indeed American lawyers have long used *Shepard's Citations* (1871–). The first *SCI* covered 1961 but only became a regular quarterly publication with an annual cumulation incorporated in the October/December issue from 1964. In 1979 it changed to bi-monthly publication with the December issue as the annual cumulation. The *Science Citation Index* consists of three separate but related indexes, each covering the same papers but indexed from a different viewpoint. The three indexes are:

(1) The Citation Index. This is an alphabetical index of authors cited. Given an author, known to have written papers relevant to one's research interests, the citation index is entered alphabetically and the author's name located; listed below his name is a list of authors and sources citing all papers of which he was the first author and which have been cited in the journals scanned by *SCI* in the period of the index. The full bibliographical reference can be located in the Source Index.

(2) The Source Index. This is a (citing) author index to over 400 000 papers taken from 2700 journals (plus 750 extra from 1979 onwards)

covered by *SCI*. A full bibliographical reference is given. This index is used to establish whether a given author has published anything over the time span of the index and to establish the full bibliographical details. It contains an index which lists organizations to which the citing authors belong.

(3) The Permuterm Subject Index. This separately purchasable index can be used when no relevant authors are known. The index is title-based; every significant word from the title is paired with every other key word. Under the significant terms is an alphabetical listing of terms linked with that term together with the author in whose title this combination occurs.

NOTE: First look under the less commonly occurring term of the pair. The full bibliographical reference can then be obtained from the Source Index. The Permuterm Index is even more tedious to use than a KWIC index and is not recommended when *Chemical Abstracts* is available.

The main method of using *Science Citation Index* is to find which articles have referred to a particular key paper during the period covered by the *Index*. This assumes that such articles will be relevant to the specific interests of the searcher. In practice a fair number of irrelevant references may be turned up. All the papers found should then be examined in the original journal if available or abstracts of them located by means of the author indexes to *Biological Abstracts, Chemical Abstracts*, etc., to find out whether the paper is relevant (these tasks can, of course, be done by relatively untrained personnel provided that they have been given an adequate set of key references to look up). This latter step applies more to the 1961 *Index* than to later issues, as these have a Source Index which gives the titles of articles and all co-authors' names. If one takes the key paper on a particular technique or analytical method as what Dr Garfield calls the target paper, one can often trace recent improvements in that technique by using the *Science Citation Index*. An example, quoted in the 1964 index, is finding an improvement on the method of Matsumoto and Sherman (1951) for the determination of mimosine, which led to the paper of Hegarty *et al.* (1964) describing an improved method. Presumably such papers (and perhaps others) could be located in due course through the subject indexes of the comprehensive abstracting services. One of the less scientific uses to which the *Science Citation Index* is doubtless put is the 'evaluation of the merit' of a particular worker's papers or merely ego gratification. Provided that these considerations do not influence authors when they are writing a paper, sinister imputations against *Science Citation Index* are probably groundless. It is interesting to note that over half the citations in the

1961 *Index* were published in the six preceding years; possibly this is due to the interdisciplinary bias of the journals scanned and, hence, the newness of many of the fields covered.

Even an experienced worker in a newly emergent facet of science is sometimes not aware of all the important primary sources. *Science Citation Index* can be used to draw up a scanning list of journals in which the majority of relevant papers are likely to occur. The scientist writes down a list of ten or twelve key references — for example, those he considers he could not possibly omit if he were writing a review of his subject of interest. These key references are looked up in the *Index* and a ranking list is drawn up of the journals in which articles appear citing the key references. As is to be expected from Bradford's Law of Scattering, a small number of journals account for over half the relevant articles, as is shown in *Table 5.1*. This technique could obviously be employed for journal selection by librarians on a tight budget in a newly founded library and to evaluate quantitatively requests for new periodical subscriptions if these are suspected of being excessive. This, of course, presupposes that the journals in question are indexed by *Science Citation Index*. This technique was first described in the second edition of this book (1969); it has many similarities to the basis of ISI's *Journal Citation Reports* service introduced in 1973, which ranks the most heavily cited journals in various fields. This is

Table 5.1 JOURNAL LIST FOR SURFACE/SOLID STATE CHEMISTRY TOPIC

C	CJ	%ΣCJ	Journals* (J)	
14	14	8.5	J APPL PHYS	
11	11	14.8	PHYS REV A	
10	10	20.5	J PHYS CH S	
8	16	30.2	BR J A PHYS	J CHEM PHYS
7	21	42.6	J ELCHEM SO	SOL ST ELEC
			J PHYS CHEM	
4	8	47.4	BER BUN GES	PHYS ST SOL
3	24	61.6	8 journals	
2	28	78.2	14 journals	
1	37	100	33 journals	
			4 US patents	
Totals	169	100	69	

Notes

C	Citations/journal (1965 and Jan–Sept 1966 *SCI*)
CJ	No. of citations in group
%ΣCJ	$= \dfrac{100 \ \Sigma CJ}{169}$
*	ISI journal abbreviations

N.B. The top 10 journals provide almost half the total references.

probably coincidental, since we have not noted any reference to it in Garfield's writings.

The reader requiring further information on citation indexing is referred to useful reviews by Garfield (1964) and by Malin (1968).

To sum up, *Science Citation Index* is probably of greatest value in an expanding interdisciplinary field, where it will perhaps be easier to write down key references rather than to list key words under which subject indexes are to be searched. It must be remembered, however, that it is not, and can presumably never be, comprehensive, in the sense that the great abstracting services are comprehensive in coverage. It is, however, a valuable second line of defence in the struggle to retrieve information, but it cannot be regarded as a means of attack superior to the subject indexes of abstracts intelligently used. Nevertheless even experienced users of *Chemical Abstracts*, etc., may locate some additional references by using *Science Citation Index*. A 52 page programmed text, *Effective Use of the Science Citation Index*, is available gratis from the ISI.

An auxiliary service is *Automatic Subject Citation Alert* (*ASCA*). Subscribers nominate 50 or more references for which the computer files are searched each week. Should any article in one of the journals processed by *Science Citation Index*'s computer cite one of these references, a print-out giving bibliographical details is sent to the subscriber, thus enabling him to learn of work which may be of interest without him actually scanning the current literature. This, of course, assumes that the subscriber has chosen his list of references wisely and also that the authors of articles of interest to him have not ignored these key references and have published in one of the journals which are indexed. Subscribers have the option to nominate authors' names, specific institutions or keywords (likely to occur in the title of an article of interest) for citations. Spelling variation can also be taken care of. Because it is a quarterly publication, *Science Citation Index* cannot be classed as an alerting service, but *ASCA* could be a most valuable SDI alerting service for anyone skilled enough to make use of it, especially for those working in an interdisciplinary field. Standard broad profiles called *ASCA Topics* are marketed; 40 of the 432 *Topics* relate to chemistry.

ISI first offered computer-readable data bases in 1966. *SCI* tapes cover a wide range of life sciences, physical sciences, social sciences and technology. There are two major components: (1) Source tapes which provide a full bibliographical record of every item from some 3700 journals covered. (2) Citation tapes. These provide a record of which authors (with full bibliographical reference) have been cited by papers from 2600 journals (plus 750 extra from 1979 onwards). The on-line searching of the SciSearch system is based on the source tapes.

ALERTING SERVICES

Several services have been started in recent years to bridge the gap between the publication of a paper and its abstract appearing in a comprehensive abstracting journal such as *Chemical Abstracts*. Mostly they are lists of article titles. Once the papers are safely garnered by *Chemical Abstracts*, these periodicals, which are designed mainly to 'increase current awareness', become of little value. The major exception to this generalization is *Current Abstracts of Chemistry & Index Chemicus*.

There are three main types of title indexes: (1) Reproduction of journals' contents pages, translated where necessary, e.g. *Current Contents*. This is the simplest and quickest to prepare and is often used by firms' libraries. (2) Classified list of titles accompanied by the bibliographical reference, e.g. the now defunct *Current Chemical Papers*. (3) KWIC Index, e.g. *Chemical Titles*. Other alerting services employ some form of SDI (Selective Dissemination of Information), e.g. *ASCA* or selective abstracts such as *CAC & Index Chemicus*. Timing tests on retrieval of information from the three types of title index showed that *Chemical Titles* was slightly quicker to use than *Current Chemical Papers* and was considerably less time-consuming than *Current Contents* (Bottle, 1965). One must, however, remember, when assessing the value and promptness of alerting services, that a service published on one side of the Atlantic and read on the other may be three or four weeks old by the time the reader sees it (unless a costly airmail edition is obtained).

Chemical Titles was one of the first products of the research division of Chemical Abstracts Service, and appears twice a month, containing about 150 000 titles per year. It is based on the tables of contents of about 700 journals of pure and applied chemistry, 110 of them Russian. Owing to the Bradford scattering effect previously mentioned, these 700 journals account for nearly half the papers abstracted by *Chemical Abstracts*. *Chemical Titles* consists of a KWIC index to the papers' titles, an author index and a bibliographical section giving the contents of each journal indexed.

In order to use *Chemical Titles* efficiently, it is first necessary to compile one's own list of keywords and their synonyms for one's field(s) of interests (as when searching the subject indexes of *Chemical Abstracts*), bearing in mind that the computer indexes words rather than concepts. The keyword index is then scanned under the headings required and the reference code noted for any article appearing with this heading which seems likely to be of interest. The reference code is then looked up in the bibliography (arranged in alphabetical order of journal codens). It is claimed that all titles appear within two weeks of their receipt by the Chemical Abstracts Service, though British readers should realize that, owing to two postal delays, they will only learn of articles in European journals some two months after their publication.

The magnetic tapes from which *Chemical Titles* is produced can be duplicated on tapes provided by subscribers who wish to make searches on their own computers. Alternatively, custom searches of a file dating back to 1962 can be made on the CAS computers or by UKCIS, etc.

Current Abstracts of Chemistry & Index Chemicus (*CAC&IC*) is a weekly publication of the Institute for Scientific Information, Philadelphia (ISI), which concentrates on new compounds. ISI claim that the 100 or so journals scanned provides a coverage of 90%. The abstracts concentrate on the graphical presentation of chemical structures and reactions. The reactions are presented by means of reaction flow diagrams, and each molecular formula is numerically labelled and its name can then be identified in the text of the abstract. Another device is the 'analytical wheel' of techniques used. Standard techniques have specific portions around the wheel together with abbreviations, e.g. NMR, MS, etc., and are shaded when used. Thus the experienced reader can tell at a glance which techniques were employed. Special techniques are indicated by an asterisk and then identified. Each issue of *CAC&IC* has: (a) a molecular formula index; (b) an author index; (c) a key word index; (d) a corporate index; and (e) an instrumental data alert where abstracts are grouped under specific types of techniques employed, e.g. NMR, thin-layer chromatography or infrared spectra. ISI now complement *CAC&IC* by *Current Chemical Reactions* (1979-), which reports novel reactions.

In addition, a monthly *Chemical Substructure Index* is available in which the compounds are encoded in Wiswesser Line Notation (WLN) (see Chapter 6). All related compounds, compounds with specific groups or derivatives appear in bands within the index owing to codes for substructures being permuted as in a KWIC index.

The bibliographical material in *CAC&IC* is available on magnetic tape, and the *Index Chemicus Registry System* (ICRS), which incorporates WLN and Ring Code tapes, can be searched using subject terms, WLN and/or ring code and any other bibliographical elements. Each monthly *ICRS* tape reports approximately 13 000 new compounds.

ANSA (*Automatic New Structure Alert*) is a computerized monthly alerting service of ISI for new organic compounds. An individual statement of interest based on chemical substructure, functional group or fragment is required. The output consists of the full bibliographical reference, author's address, analytical techniques used, any subject indexing terms applied and an abstract number referring to *Current Abstracts of Chemistry & Index Chemicus*. Retrospective searches from 1966 can be carried out.

Current Contents, Physical and Chemical Sciences is a weekly alerting service which photoreproduces the contents page of a wide range of journals (often obtained at the page proof stage). Journals, by their editorial policies, can themselves define the area of coverage, and a user is able to establish fairly quickly which are the core journals in his field. Relevant material scattered among other journals is also of interest and to aid retrieval of such articles an alphabetical keyword index has been introduced. ISI produces a family of *Current Contents* with partially overlapping journal coverage. Each is provided with an abbreviated author address directory.

Since the ISI air-freights its publications to Holland before posting to European subscribers, transatlantic postal delays are considerably reduced and thus *Current Contents* normally notifies us of a paper a little before *Chemical Titles.*

Chemischer Informationsdienst (Chem. Inform.) commenced when *Chemisches Zentralblatt* ceased publication in 1969. *Chem. Inform.*, a combination of the *Fortschrittsberichte* of the BAYER company and the *Schnellreferate CZ* of Gesellschaft Deutscher Chemiker, is produced jointly by these two organizations and published by Verlag Chemie. *Chem. Inform.* is published weekly, its aim being current awareness. The journal was published in two parts covering (1) physical and inorganic chemistry, and (2) organic chemistry. However, after 1973 these two issues were combined into one journal. Approximately 350 journals are covered, and selection of articles for inclusion is made by a panel from industrial and academic institutions. In the weekly issues the abstracts are ordered according to a classification system which consists of five main parts each with further sub-sections. The five main parts are: general subjects; physical-inorganic and inorganic chemistry; physical-organic and organic chemistry; applied chemistry; and analysis. Standardized keyword indexes and author indexes are produced at the same time as the abstracts journal. Each weekly issue carries 600—700 abstracts. The aim of the service is that of current awareness rather than complete documentation. Core journals are fully covered, while selection from fringe journals is carried out by industrial and academic experts. Graphical representation of chemical reactions is emphasized in the abstracts. In addition, a card file of 'trivial names' is produced. This correlates English and German trivial names and links them to the molecular and structural formulae of the compound.

The American services are increasingly expensive, and some users will doubtless prefer the personalized Current Awareness Services supplied by the Scientific Documentation Centre, Dunfermline, Scotland. Each week subscribers receive references on either 5 X 3 inch or 80 column machine-sortable cards selected from over 2500 journals and US Government Research and Development Reports and US and British thesis titles. Currently about 200 topics are covered, but the Centre is prepared to quote for other topics on demand. Retrospective searching facilities are available for many of the topics.

The titles of papers accepted for publication are sometimes available several months before their actual appearance, e.g. in the *Biochemical Journal*. Sometimes they are listed in a society's news journal; sometimes authors' summaries are actually sold separately, as in the case of *Biochimica et Biophysica Acta Previews*. They raise an interesting question — are such publications primary or secondary literature? They almost come under the heading of Research Intelligence, another facet of current awareness, which is discussed in Chapter 16.

ON-LINE SERVICES AND NETWORKS

Electronic communication is gaining momentum and, with the development of such systems as BBC's *Ceefax* and the Post Office's *Prestel* which are intended for business and home use, the scientific community has to become geared to the use of computerized retrieval systems. It should be pointed out that with the exception of complex topics it is cheaper to carry out full retrospective searches using hard copy. Moreover, most data bases only cover quite recent material. For example, MEDLARS/MEDLINE coverage is available from 1964, while *Chemical Abstracts* is only from 1969.

The main advantages of searching on-line are, firstly, that recent material is available more quickly than it is in hard copy and, secondly, that the heuristic nature of the interaction between searcher and data base enables the searcher to alter the search strategy according to the nature of the output. There is, however, no standardization of input. Consequently, in the same system, the searching of data bases may be based on thesaurus-controlled indexing terms, while in other cases searching for the same topic may be based on 'free-text' abstracts. The detailed consideration of such systems is beyond the scope of this chapter, but the coverage of secondary services would not be complete without mentioning them.

Several computer-readable data bases have already been mentioned. A number of these are made available on-line. There are a number of systems operating on-line services. The major services available are:

Lockheed **DIALOG**. The main computer is situated in Palo Alto, California, with centres in Europe and the USA. Numerous data bases are available covering a wide range of subject fields. Those relevant to the chemical field include *BIOSIS, CA Condensates, CA Patent Concordance, Chemical and Chemically Related Patents, Engineering Index COMPENDEX*.

The Space Documentation Service of the European Space Agency — RECON. The main computer is situated in Frascati, Italy, but is accessed through local centres throughout Europe. The UK centre for SDS/RECON is at the Technology Reports Centre, St Mary Cray, Kent; in the UK the service is known as DIAL-TECH. The data bases relevant to the chemical field are *CA Condensates, INSPEC, COMPEN-DEX, Metals Abstracts, USGRDR* and *Environmental Index*. Printing facilities for output of searches are available.

System Development Corporation's SDC/ORBIT is located in Santa Monica, California, and available through the Tymnet network. There are some 30 data bases available. Those that are relevant to the chemical field are: *CA Condensates; BIOSIS; Smithsonian Scientific Information Exchange (SSIE)*, which gives fairly comprehensive coverage of government and foundation research and is particularly good on medicine, agriculture and energy; *API/LIT* and *API/PAT*, from American Petroleum Institute, which concentrate on petroleum and petroleum products literature and patents; *TULSA*, produced by the University of Tulsa and concerned with oil and gas exploration; Derwent's *World Patent Index (WPI)* and *RINGDOC*; *PAPER CHEM*, produced by the Institute of Paper; *ENERGYLINE*, from *Petroleum Energy News*, which concentrates on new fuels; and *Dissertation Index*, based on *Dissertation Abstracts*.

EURONET, the European On-line Information Network, is now being implemented by the Commission of the European Communities. It is hoped that by mid-1979 some 100 data bases will be available. These will consist of the major services from member countries and others. The *Chemical Abstracts* data base is expected to be available via a new British service, INFO-LINE, which will be linked to EURONET.

BLAISE (British Library Automated Information Service) is an on-line service which began in the spring of 1977. MEDLINE, together with CHEMLINE and TOXLINE, were the first to be made available, followed by both UK and Library of Congress MARC tapes on which English-language and foreign material is on two separate files which can be searched on a wide range of data fields.

KEEPING PERSONAL RECORDS

In the course of his career the chemist will read many papers (and learn of many more) which may be relevant to his present or projected work. All these should be noted down before they are forgotten and in such a manner that their essential points can be retrieved. Personal idiosyncrasies will doubtless dictate the pattern of this; nevertheless, a few hints for guidance are proffered.

In such personal abstracts many abbreviations can be used which would be unacceptable in published abstracts. It is expedient, however, to follow the layout suggested on pages 63—64. It is helpful to write the key words in the title in capitals and also to underline the authors' names, volume number and/or year of publication, to make these stand out. The question now arises as to whether to write them in a book or on cards. It is adequate to write the abstract in a stiff-covered book, provided that the abstracts are sequentially numbered and that the document number is punched on to all the appropriate (body-punched) descriptor cards. This may seem like cracking the problem with the proverbial sledge-hammer, but in the authors' experience it does not take many years before abstracts written even in edge-indexed books take too long to search out (assuming one still remembers that they exist).

The use of punched cards for recording literature references by the chemist *from the beginning* of his career cannot be sufficiently emphasized; otherwise much of his early library work will tend to become valueless. He must, however, lay down a systematic and expandable code for punching at the start.

One of the most convenient methods is to record the reference and abstract on an edge-punched card. The concepts covered by the article and authors are coded into the pattern of holes notched.

If punched cards are not available, a hundred or so references can be dealt with comfortably by using plain, 5 × 3 inch cards, filed alphabetically between stiff separator cards with projecting tags, especially if set out like the examples in *Figures 5.1* and *5.2*.

Only very brief abstracts were made in these two cases as the originals were readily available. Certain experimental details, not normally given in abstracts, were recorded for quick reference during

---

The Continuous ELECTROPHORESIS of Wheat GLUTEN

H. Zentner, Chem. & Ind., 1960, 317-8 (19/3/60)

Wet gluten dispersed in 0.01m lactic acid.
1340V., 16ma., 24 drip points.
Stained curtain showed 7 protein fractions.
Main fraction = 54% (no. 2) is lipoprotein
(stained with Oil Red O) + some unidentified
CHO.

(cf. Idem ibid. 1958, 129.)

Figure 5.1

```
Fractionation of GUM ARABIC by Chemical &
IMMUNOLOGICAL procedures.

M. Heidelberger, J. Adams & Z. Dische,
J.A.C.S., 1956, 78, 2853-5.

GA from specific ppt. with Type II anti-
pneumococcus serum contains only 0.3-0.2 rhamnose
of O.S. (Analyses by Dische's method, q.v.)
Up to 2% GA solns. in 0.9% NaCl, insol. material
centrifuged off.
```

Figure 5.2

experimental work on the problems in hand. Only the first letters of words of minor importance for indexing in the title are written in capitals (e.g. Continuous and Wheat in *Figure 5.1*). Where there are several authors, the chief author's name is doubly underlined. As can be seen, non-standard abbreviations are often used.

REFERENCES

Anon. (1967). *CAS Today: 60th Anniversary Edition*, Chemical Abstracts Service, Columbus
Atsu, A.S.K. (1976). *M.Sc. Thesis*, The City University
Bottle, R.T. (1965). *J. Docum.*, 21, 177
Brookes, B.C. (1977). *J. Docum.*, 33, 180
Cahn, R.S. (1965). *Survey of Chemical Publications and Report to the Chemical Society*, Chemical Society, London
Garfield, E. (1964). *Science*, 144, 649
Hegarty *et al.* (1964). *Aust. J. Agric. Res.*, 15, 168
Hill, E.A. (1900). *J. Amer. Chem. Soc.*, 22, 478
Malin, M.V. (1968). *Library Trends*, 16, 374
Matsumoto, H. and Sherman, G.D. (1951). *Arch. Biochem.*, 33, 195

PROOF ADDITION: ON-LINE SERVICES AND NETWORKS

A new chemical information system, CIS, owned by the National Institutes of Health/Environment Protection Agency (NIH/EPA) is being launched by UKCIS in the UK. In Europe it is accessible on-line via the Telenet data communications network. The System comprises several chemical data banks and bases and programs for manipulating data. Coverage includes information relating to oil and hazardous materials, mass spectra, X-ray diffraction, acute toxicity, NMR and the *Merck Index*. The data can be accessed by chemical structure via *CA* Registry Numbers.

6

Nomenclature, compound indexes and chemical coding

R. T. Bottle and *C. Oppenheim*

ORGANIC CHEMICAL NOMENCLATURE

A basic problem in the retrieval of information about an organic compound is the need for an assurance that a search has been made under the 'correct' name of the substance as defined by the tool being searched through. Most, if not all, organic compounds can be validly named in more than one way, and thus a compound may appear under several different names in various sources or indeed in the same source. Similarly, when a compound is met for the first time by a secondary service, a name must be devised for the new structure.

Ideally, then, there should be one strict set of rules for the nomenclature of organic compounds to which all would adhere, and indeed considerable progress has been made in recent years to establish such a systematic set of rules. These rules were drawn up by International Union of Pure and Applied Chemistry in 1957, and are normally referred to as the IUPAC rules. Formerly published by Butterworths, the latest version is *IUPAC: Nomenclature of Organic Chemistry* (Pergamon Press, 1979). The nomenclature for types of compounds not yet covered by definitive IUPAC rules are studied by committees, who publish tentative rules for the naming of such compounds, often in *Journal of Chemical Information and Computer Sciences*.

Unfortunately, the IUPAC rules often still allow more than one possible name for a given compound. Thus, for example, a simple molecule (*Figure 6.1*) can be named as follows under IUPAC rules:

Figure 6.1

4-carboxy-3-hydroxy-2-phenylquinoline
3-hydroxy-2-phenyl-4-quinolinecarboxylic acid
4-carboxy-2-phenyl-3-quinolol

As *Chemical Abstracts* (*CA*) is the most widely used literature-searching tool in chemistry, its policy on the nomenclature of compounds in its indexes is of great influence. Over the years, *CA* has consistently moved towards a more systematic nomenclature, so that now only a handful of trivial names are permitted. (A search for a particular compound through *CA* subject indexes over the years is therefore complicated by the need to be aware of when compound names were changed by *CA*. Fortunately, *CA* policy is outlined regularly in its *Index Guides* or in the Introductions to the Subject Indexes.)

The *CA* rules are based on those of IUPAC. However, the IUPAC rules have been extended by *CA* to eliminate choices of names. This is necessary for consistency in *CA* subject indexes.

Mention should also be made of the CAS Registry System. CAS staff perfected an algorithm for generating a unique and unambiguous computer-language description of a chemical substance's structure. The Registry System employs this algorithm to assign each chemical structure a numerical identifier (CAS Registry Number) which designates one substance only and which can be directly linked to the *CA* name. Thus the *CA* name for any substance which has already been inputted into *CA* can be quickly obtained from its structure. This saves considerable indexer time and effort in naming compounds.

The following are recommended as guides to organic nomenclature:

IUPAC: *Nomenclature of Organic Chemistry* (2nd edn, Butterworths, 1971)
An Introduction to Chemical Nomenclature (R.S. Cahn, 5th edn, Butterworths, 1979)
The Names and Structures of Organic Compounds (O. T. Benfrey, Wiley, 1966). A programmed text which is suitable for non-chemists.
Nomenclature of Organic Chemistry (Pergamon Press, 1979). This is the 1979 edition of the 'Blue Book' containing detailed IUPAC recommendations.

Compounds of recognized pharmaceutical importance often have an officially sanctioned contracted name (approved name) which is listed in *USAN,** *Approved Names*† or in the Merck Index (page 106) (e.g. prednisone for 17α,21-dihydroxypregna-1,4-diene-3,11,20-trione). These names do not normally convey as much information or produce as quick a sense of unambiguous recognition of the specific compound involved as does even the IUPAC name, let alone the structural formula.

A machine-readable file (*CIMI-Drugfile*) of some 5000 chemicals used in pharmaceutical products gives the generic or established chemical name, Wiswesser Line Notation, *CA* Registry Number and claimed pharmacological action. Thus all compounds with particular structural features or therapeutic applications can be retrieved from the file, which is available from Chemical Information Management Inc., Cherry Hill, NJ.

Trade names create similar problems. Some sources of information are given on page 231.

Inversion in name indexes

One of the characteristics of name indexes in organic chemistry is the use of inverted forms of the chemical names, e.g.

    acetic acid
    acetic acid, trichloro
    acetic acid, trichloro, methyl ester

A portion of the compound is selected on the basis of its indexing value and is named as the 'parent compound'. All remaining portions are then named as substituents. The purpose of inversion is to bring together related compounds in the index, but inevitably not all related compounds will be drawn together.

FORMULA INDEXES

Formula indexes have become very important in recent years because of the complexity of chemical nomenclature, which has, in the words of one author, 'reduced chemists to the status of laymen when it comes to searching the literature of their own field'. Even if a chemist cannot be sure of the 'correct' name for a compound as used by the searching tool he is employing, there can rarely be any doubt of its empirical

---

**USAN and the USP Dictionary of Drug Names* (13th edn, 1961–75, US Pharmacopeial Commission, Rockville, Md.).
†British Pharmacopoeia Commission *Approved Names* (HMSO, rev. edn, 1977–), which is kept updated by supplements.

formula. In most formula indexes the names of all the relevant compounds are listed against the formula, and it is much easier to select the name from this list which fits the structural formula than to write down the name used by the indexers. Information on the compound can then be obtained from the name index or the information may be directly linked to the formula index, e.g. in *CA*.

There are two major systems for ordering the formulae: the Hill and the Richter systems. In the Hill system the formulae are arranged in alphabetical order of symbols, except that in carbon compounds C comes first, followed immediately by H if it is present, and then by the other symbols. As in the Richter system, all compounds containing one carbon atom are listed before any C_2 compounds, and so on. The Hill system, which can cater for inorganic as well as organic compounds, was devised in the 1900s for use in the United States Patent Office. It is used by *CA, Beilstein* (for indexes of *Supplement 2*) and the *Handbook of Chemistry and Physics*, and is by far the more popular of the two systems. Further details are on page 70.

The Richter system, which caters for organic compounds only, was used in earlier parts of *Beilstein*, and is still used in *Lange's Handbook of Chemistry*. Elements are arranged in the order C, H, O, N, Cl, Br, I, F, S, P, followed by the remaining elements in alphabetical order of symbol.

The example in *Figure 6.2* shows the differences between the two systems.

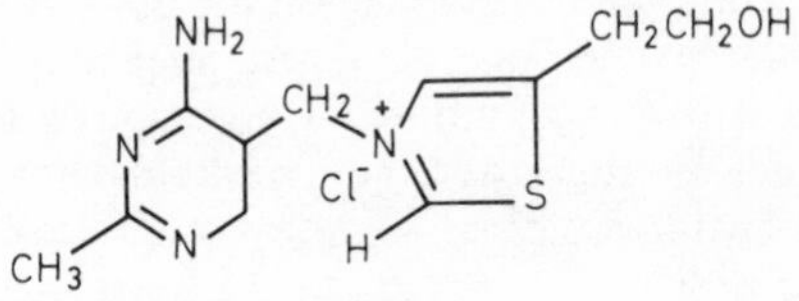

Thiamine

$C_{12}H_{17}ClN_4OS$ (Hill)
$C_{12}H_{17}ON_4ClS$ (Richter)

Figure 6.2

Formula indexes can be arranged in a manner analogous to KWIC and KWOC indexes. In this way, particular atoms (other than C and H) can be prominently displayed. Examples of these sorts of indexes are *CA*'s short-lived HAIC (Hetero-atom in context), and the 'Rotaform Index' to *Current Abstracts of Chemistry and Index Chemicus* (*CAC& IC*). Descriptions of both these indexes appear in the introductions to *CA* and *CAC&IC* respectively. The 'Rotaform Index' appears only in the quarterly and annual cumulated indexes to *CAC&IC* and has in many respects been overtaken by the *Chemical Substructure Index*, a Wiswesser Line Notation (WLN) Index to the compounds disclosed in *CAC&IC*.

RING INDEXES

Because of the importance, variety and complexity of ring systems, special indexes based on their skeletal structure have been developed. The best-known are *The Ring Index* (A. M. Patterson, L. T. Cappell and D. F. Walker, 2nd edn, ACS, 1960, + 3 supplements) and the *Index to Ring Systems* which forms part of the indexes to *CA*. Both employ a special code based on the cyclic structure of the ring, regardless of the presence of hydrogen atoms or substituents. First, the systems are arranged in numerical order according to the number of rings present. This number is calculated by determining the number of scissions needed to convert the ring structure to an open-chain system. Within this arrangement the rings are arranged in increasing ring size. Within this they are arranged by formula of each component ring, using the Hill system. Finally, the chemical names (and sometimes structural formula) are displayed. The *Ring Index* provides one early reference to the ring system, while the *Index to Ring Systems* gives the *CA* name under which the item can be found in the *Chemical Substance Index*.

Both the *Ring Index* and the *Index to Ring Systems* only consider single and fused ring systems. Thus the thiamine system (*Figure 6.2*) is not given a separate entry, although the thiazole (C_3NS) and pyrimidine (C_4N_2) rings are to be found separately. Users should also note that the second edition of *The Ring Index*, with its supplements, covers only the chemical literature between 1907 and 1963. Ring systems discovered since then are to be found in the *Index to Ring Systems*.

The *CA Index Guides* are useful for the reverse information to that in the *Index to Ring Systems*, i.e. for giving the structural diagrams, properly numbered, which correspond to a given name. In 1968 the ACS published *Wiswesser Line Notations Corresponding to Ring Index Structures* (PB 180, 901).

The *Ring Index* and its *Supplements* have now been replaced by the *Parent Compound Handbook*. The *Handbook* consists of a Parent Compound File and an *Index of Parent Compounds*. Bi-monthly loose leaf supplements are issued for both parts and the Index will be reissued biennially. The File gives structural and molecular formulae, Registry Numbers, WLN, the current *CA* Index Name and post-1972 *CA* references for each parent compound. The *Handbook* additionally contains some 2000 cyclic and acyclic natural product stereochemical parent compounds and about 100 polyboranes, metallocenes and ferrocenophanes. It is valuable for finding the *CA* Index Name and/or WLN for such compounds and other ring systems prior to searching *CA* indexes, *ICRS*, etc.

CHEMICAL CODING

One of the major problems with organic nomenclature which especially applies to trivial names is their dependence on language. Thus, it would be difficult to predict that the German for succinic acid would be 'Bernsteinsäure', and for urea would be 'Harnstoff'. The same problem applies to a lesser extent in systematic names; for example, the German for quinoline is 'Chinolin' and even the suffix for normal paraffins changes from 'ane' in English to 'an' in German.

Formulae indexes are not language-dependent, but suffer from the disadvantage that any one molecular formula can represent many different isomers.

Structural formulae are independent of language (even in Japanese) and, excluding stereoisomers, are unambiguous. They convey much information to the chemist — far more (and more quickly) than a systematic name can. However, structural formulae, being two-dimensional representations of a normally three-dimensional molecule, are difficult to input into, or output from, a computer, since computers require a linear sequence of input or output signals.

The Beilstein system (Chapter 11) provides a method for arranging all organic compounds sequentially according to structural criteria, but is not adaptable to mechanized handling. For optimum automation of a storage and retrieval system based on structure, it is essential to convert the structures to a linear sequence of symbols. These can then be printed out or stored on magnetic tape or disc. Such a notational system must also be independent of nomenclature and preferably be amenable to substructure searching.

Three groups of such coding systems are used in chemical information systems — fragmentation codes, linear notations and topological codes.

Fragmentation codes can be viewed as a development of the traditional method of classification by type or functional group, e.g. esters, ketones, etc. A compound is represented as a composite of its predominant structural features. Many codes have been designed to facilitate correlations between structure and biological activity.

Linear notations are relatively compact ordered strings of alphabetical and numerical symbols used to represent chemical structures. They were intended as alternatives to nomenclature, using a smaller vocabulary of symbols and a simpler syntax, and therefore easier to apply and interpret.

The availability of punched card equipment influenced the development of fragmentation codes and linear notations. In the same way the increasing availability and acceptance of the computer as an information-handling machine opened up further possibilities, including the introduction of topological codes. These codes are structure descriptions showing atom-to-atom connections, with the atoms as nodes and the connections as branches of a network. The CAS Registry System is the best-known example. Rush (1976) has reviewed the present status of notation and topological systems and potential future trends.

LINEAR NOTATION OF CHEMICAL STRUCTURE

Because molecular formulae and many names of organic compounds do not reveal structure, and as noted earlier, nomenclature is not completely systematic, both have serious deficiencies as indexing tools. Several attempts have therefore been made to develop linear notations to represent the structures of chemical compounds by unique and unambiguous sequences of letters, numbers or other symbols. Such notations may be regarded as an algebra which reflects the topological properties of a chemical structure.

The optimum linear notation system will assign to every chemical compound a unique place in an order list of notations, and this list will therefore be an index. The most useful characteristic of a notation is its potential to serve as a superior indexing tool and one which is much more adaptable to mechanized manipulation than existing schemes.

The key idea of structural chemistry was developed at the first International Chemical Congress at Karlsruhe in 1860. Shortly after, in 1861, A. M. Butlerov used the term 'structure' in its modern chemical meaning for the first time. Within the next seven years a small group of workers developed almost all of the important concepts of line-formula descriptions (Wiswesser, 1968).

The older organic chemistry texts abound with formulae such as: $CH_3CH_2CH_2OH$, $C_3H_7.O.C_2H_5$, $p\text{-}HOC_6H_4NO_2$, $Me.CO.Et$, $PhNH_2$, etc. Wickelhaus in 1867 used dots and dashes to show side and main chain connections, e.g. $CO.OH-CH.OH-CO.OH$. Since that time several line-formula notations have been developed.

Apart from the use of Me, Et, Pr, Ph, 'R', etc., no major development took place until 1944, when Dyson devised a system for a unique and unambiguous 'ciphering' of chemical structures built up logically from first principles, which, unlike the IUPAC systematic nomenclature, was uncluttered with ghosts of trivial names from the past.

Two contrasting approaches have been used for linear notations:

(1) Hierarchical order (as with systematic nomenclature); cf. Dyson.
(2) Sequential order; cf. Wiswesser.

Three linear notations (those of Dyson, Hayward and Wiswesser) have been carried to an advanced state of development. Although IUPAC provisionally accepted the Dyson notation in 1951, only the Wiswesser notation has received significant use in operational systems for the storage and retrieval of information about chemical structures. The Dyson scheme was seriously considered by Chemical Abstracts Service, though in the end the Registry Number system, a topological code, was used instead.

A 1976 census revealed that nearly 800 000 Wiswesser notations had been entered into a total of fourteen operational or experimental storage and retrieval systems in the US and the UK. Since 1970 the Wiswesser line notation has been employed by the Institute for Scientific Information to encode approximately 150 000 new compounds a year that are reported in *CAC&IC*, and are machine-searchable through its *Index Chemicus Registry System*. Some 1.5 million compounds are now in the *ICRS*.

WISWESSER LINE-FORMULA CHEMICAL NOTATION

The Wiswesser Line-Formula Notation expresses the structural formula of chemical compounds concisely by using letters to denote functional groups and numbers to express the lengths of alkyl chains and the sizes of rings. These symbols are written in a specified connecting order from one end of the molecule to the other.

The 41 symbols used are the 10 numerals, 26 capital letters, 4 other symbols (&, −, / and *) and the blank space, since these symbols are all available on existing computers and punched card accounting machines, and their use makes possible a variety of mechanized sorting, searching, list-printing and retrieval procedures for chemical structures with existing machines.

Since the symbols for functional groups or non-benzenoid rings are frequently starting points of the notation, the Wiswesser Line-Formula Notation (WLN) focuses on structurally significant features of compounds. Alphabetical arrangement of the notation has a strong tendency to bring together compounds of related structure, and this makes feasible the creation of WLN indexes amenable to substructure

Table 6.1 MAJOR WISWESSER LINE-FORMULA NOTATION SYMBOLS

All the international atomic symbols are used except K, U, V, W, Y, Cl and Br. Two-letter atomic symbols in organic notations are enclosed between hyphens. Single letters preceded by a blank space indicate ring positions. Single letters not preceded by a blank space have the following meanings:

C unbranched carbon atom multiply bonded to an atom other than carbon (−CN, −SCN, etc.) or doubly bonded to two other carbon atoms
E bromine atom
F fluorine atom
G chlorine atom
H hydrogen atom where expressed, or when preceded by a locant within ring signs, shows the position of a carbon atom bonded to four other atoms
I iodine atom
J generic halogen; ring closure symbol
K nitrogen atom bonded to more than three other atoms
L first symbol of a carbocyclic ring notation
M −NH− group
N nitrogen atom, hydrogen-free, attached to no more than three other atoms
O oxygen atom, hydrogen-free
Q −OH group
R benzene ring
T first symbol of a heterocyclic ring notation; or, within ring signs, indicates a ring containing two or more carbon atoms each bonded to four other atoms
U double bond (but not carbonyl)
UU triple bond
V carbonyl connective, $-\overset{O}{\underset{\|}{C}}-$ (carbon atom attached to oxygen and two other atoms)
W non-linear (branching) dioxygen group, as in $-NO_2$ or $-SO_2-$
X carbon atom attached to four atoms other than hydrogen
Y carbon atom attached to three atoms other than hydrogen or doubly bonded oxygen
Z −NH₂ group

NOTE: Numerals not preceded by a space show ring sizes if within ring signs; elsewhere numerals show the length of internally saturated, unbranched alkyl chains and segments.

searches. The notation may be read, or decoded, at sight, like a conventional structural formula, by any chemist who will learn about a dozen chemical symbols to supplement the old familiar ones, plus a few operating symbols and the rules for manipulating them. At the present stage of development there are 89 such rules plus a few more for dealing with polymers. The development of WLN is now in the hands of the Chemical Notation Association, and their definitive text is *The Wiswesser Line-Formula Notation (WLN)* (E. G. Smith and P. A. Baker, Eds, 3rd edn, Chemical Information Management Inc., Cherry Hill, NJ, 1975).

The most frequently used rules could probably be learned in several days of intensive effort by any organic chemist. We will only look at five of these to illustrate the technique of coding by WLN.

Rule 1 Cite all chains of structural units symbol by symbol as connected (see *Table 6.1*).

Rule 2 Resolve all otherwise equal alternatives in symbol sequences by selecting the sequence that would give the notation the *latest* position in an alphanumeric list arranged in the sequence Blank & − / 0 1 2 3 ... 7 8 9 10 11 12 (etc.) A B C ... X Y Z * (but note Rule 20).

Examples

1. $CH_3SCH_2CH_2CH_3$ 3S1

2. $CH_3CH_2\overset{\overset{O}{\|}}{C}OH$ QV2

3. $ClCH_2\overset{\overset{O}{\|}}{C}OH$ QV1G

4. $H_2NCH_2\overset{\overset{O}{\|}}{C}OH$ Z1VQ

5. $CH_3CH_2\overset{\overset{O}{\|}}{C}OCH_2CH_3$ 2VO2

6. $CH_3S\overset{\overset{O}{\|}}{C}NHCH_3$ 1SVM1

7. $BrCH_2CH_2Br$ E2E

Rule 3 Use U to denote a double bond in a chain except >C=O (V). Use UU for an acetylenic bond.

Examples

8. $CH_3CH=CHCH_3$ 2U2

9. $CH_3CH=NOH$ QNU2

10. $CH_3CH_2CH_2OCH=NH$ MU103

11. $CH_3CH_2\overset{\overset{S}{\|}}{C}H$ SU3

12. $HC\equiv CCH_3$ 2UU1

13. $CH_3C\equiv CCl$ G1UU2

Then follow rules for dealing with branched chains, etc., and various permitted contractions.

Rule 20 Subordinate the benzene symbol R to all other symbols by ranking it in earlier than & (exception to Rule 2).

Rule 21 Denote disubstituted benzene compounds by the following steps:

(a) Ignoring ring positions write out the notation oriented by Rules 2 and 20.
(b) In the structural formula, assign the 'a' locant to the ring position of the substituent that precedes R in this order.
(c) Assign consecutive letter locants to consecutive ring atoms in the direction that will give the *earliest* locant for the second substituent.
(d) Into the oriented notation produced in step (a) insert immediately before the second substituent symbol a blank and the ring locant letter B, C or D determined in step (c). NOTE: In handwritten WLNs the blank is often emphasized deliberately by underlining it, e.g. Ex. 14. ZVMR_CG.)

Examples

| | Step (a) | Steps (b) & (c) | Step (d) (complete WLN) |

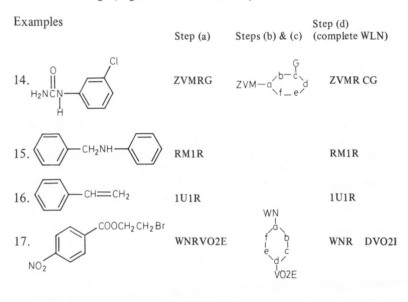

Homocyclic rings

Homocyclic rings other than benzene commence with L and the notation ends with J. The number of atoms in each ring is given and then the position of ring segments and substituents is indicated by locants, as are ring fusion points in polycyclic compounds.

Heterocyclic rings

Heterocyclic rings commence with T and indicate the heteroatom after the ring size, other heteroatoms are noted and positioned by locants and the ring is closed by J.

The rules are complex, but some examples may suffice to indicate the type of notation encountered:

L6V DVJ   T6NJ   T6M CMTJ   T6NJ CVQ DG   L66J CVQ

WLN lends itself to substructure searching, and several KWIC indexing programs have been written for it, including Double KWIC. One of the best-known programs is CROSSBOW (Computerised Retrieval of Organic Structures Based On Wiswesser), originated by ICI, which converts WLNs to structural formulae. Polymers were listed as a problem area in the second edition of Smith's book but the Chemical Notation Association's Polymer Committee Report (September 1971) gives additional rules for handling linear polymers. Polymers are readily recognizable, as they have the only notations beginning with /. The Report is summarized by P. A. Baker, G. Palmer and P. W. L. Nichols in *Chemical Information Systems* (J. E. Ash and E. Hyde, Eds, Ellis Horwood/Wiley, 1975).

IUPAC/DYSON NOTATION

This notation has been described in detail in *Rules for the IUPAC Notation for Organic Compounds* (Longmans, 1961), and a more recent modified notation was described by Dyson and co-workers (1968). Both notations divide the structures of compounds into carbon chains, substituents on them and heteroatomic links between them. In general, conventional symbols are used for elements. Cyclic compounds are designated either A (saturated) or B (aromatic). Heterocyclic structures are indicated by the symbol Z followed by the symbol for the element.

Table 6.2 COMPARISON OF DYSON AND WISWESSER NOTATIONS
(FROM HORSFALL, 1974)

Structural formula	IUPAC/Dyson	WLN
C_2H_6	C_2	2H
CH_3OH	CQ	Q1
C_2H_5OH	C_2Q	Q2
(benzene ring)	B6	R
(phenol, OH)	B6Q	QR
CO_2H / CH_2 / OH	C_2X1Q_2	QV1Q
CO_2H / $(CH_2)_2$ / OH	C_3X1Q_2	QV2Q
(benzene ring with CO_2H and OH)	$B6X1Q_4$	QVR DQ
(structure with C_2H_5, Me, HO, NH_2, CO_2H)	$C_9C_2$4C6E2YZ9Q1N5	QV1UU1Y&Y2Y2&1U2Q
(structure with C_2H_5, phenyl, HO, NH_2, CO_2H)	B6:C/6C9C 4E2YZX9Q1NS	QV1UU1Y1R&Y2Y2&1U2Q
(pyridine ring, N)	B6ZN	T6NJ

Table 6.2 (*continued*)

	A5ZQ	T5OTJ
	B65ZQ7	T56 BOJ
	B65617EQ9	LB656 HVJ

Finally, certain common functional groups have special symbols. *Table 6.2* compares the IUPAC/Dyson notation and the Wiswesser Line Notation for a selection of organic compounds.

HAYWARD LINEAR NOTATION

Hayward developed a notational system for organizing the chemical structures on file at the United States Patent Office, and the notation is specifically designed for dealing with general or Markush-type structural formulae covering many different compounds. The notation is described in detail in *United States Patent Office R & D Report* No. 21 (1961).

ACKNOWLEDGEMENTS

We should like to thank Mrs Christine Horsfall for permission to reproduce *Table 6.2*.

REFERENCES

Dyson, G.M., Lynch, M.F. and Morgan, H.L. (1968). *Inf. Storage Retr.*, **4**, 27
Horsfall, C. (1974). *M.Sc. Thesis*, The City University
Rush, J.E. (1976). *J. Chem. Inf. Comp. Sci.*, **16**, 202
Wiswesser, W.J. (1968). *J. Chem. Docum.*, **8**, 146

7

Background information and ancillary literature keys

R. T. Bottle

Most of us will want, at some time or other, to fill in the background knowledge between what we already know and what must be known about a particular field in order profitably to work in it and/or peruse its specialized literature. Starting from scratch, a possible sequence is (1) acquiring terminology and an elementary conspectus of the field through dictionaries and encyclopaedias, (2) acquiring a basic grasp of the topic through textbooks, monographs and reviews aimed at the non-specialist, (3) reading specialized reviews and/or critical surveys by leaders in the field (4) obtaining a key to the literature. Frequently one does not need to go beyond the first stage of looking in dictionaries and encyclopaedias for answers to one's shorter queries. In recommending sources of background information one must bear in mind that availability is almost as important as reliability. Even quite small public libraries have a recent edition of the *Encyclopaedia Britannica*, which is an often neglected source for one's initial reading, even if not a particularly up-to-date one. The other more specialist encyclopaedias referred to later may not, however, be so readily available.

The obvious source of background information is the advanced textbook, which, as its compass becomes smaller and its treatment more comprehensive, tends to be called a monograph. The review article is becoming an increasingly prevalent and valuable source of chemical information, and can often be the key to much of the original literature on a subject. Theses and dissertations are, or should be, critical reviews as well as a much neglected source of primary information, listing experimental details and pitfalls which may well not be published elsewhere, and thus are discussed in Chapter 3. The boundary between the non-critical review and the annotated bibliography is rather vague. Often reviews are sought as a short-cut to compiling one's own bibliography. A number of reviews are to be found in papers presented at conferences. In spite of electronically produced journals, some chemists feel that the best way of getting the latest information is by word of mouth from the experts through attending symposia, etc., as did the ancient Greeks. While this may not be possible for us all, it should not be forgotten that published proceedings of symposia, conferences, etc., often form a useful source of background information in a particular field.

The complete background to a problem includes a knowledge of *who* is doing *what* and *where*. This will doubtless be known (or thought to be known) by the experienced worker, but such information can be obtained by the new graduate from the literature. Background information of this type is, however, more appropriately discussed along with other less conventional sources of information in Chapter 16. Encyclopaedias, monographs, reviews, bibliographies and conference literature form natural divisions of this chapter. In order to try to keep it readable, exhaustive lists of books, journals, etc., are not given; only illustrative examples and instructions for locating further ones are given.

CHEMICAL ENCYCLOPAEDIAS, DICTIONARIES AND OTHER QUICK REFERENCE SOURCES

When using general encyclopaedias, one should check the date of the most recent reference to judge just how current the information is. This is particularly important with *Britannica*, which has recently been revamped into three parts, the *Propedia* (one volume), *Macropedia* (ten volumes), and *Micropedia* (ten volumes), the last also acting as the index to the *Macropedia*. For example, in the 1975 edition the most recent reference to transition metals was 1972 and to uranium was 1969.

Apart from articles in general encyclopaedias, there are several encyclopaedias devoted to chemistry and its technology, which will often provide an answer to one's shorter queries. The most useful of these is Kirk and Othmer's *Encyclopaedia of Chemical Technology* (Interscience, 2nd edn, 22 vols, 1963–70 + supplement, 1971 and Index, 1972; 3rd edn, 25 vols, 1978–). The German equivalent which currently updates *Kirk–Othmer* is the fourth edition of Ullmann's *Encyclopädie der technischen Chemie* (23 vols, E. Bartolomé, Ed., Verlag Chemie, 1972–). Slightly more up to date than *Kirk–Othmer*'s second edition is the eight-volume *Materials and Technology* (Longmans, 1968–75), published in the USA as *Chemical Technology* by Barnes and Noble. This is based on the sixth edition of *Warenkennis en Technologie* (J. W. van Oss, Ed., de Bussey, Amsterdam). The enquiry service for users of this encyclopaedia is described in the preface to each volume. The volumes are:

1. Air, Water, Inorganic Chemicals and Nucleonics, 1968
2. Non-metallic Minerals and Rocks, 1971
3. Metals and Ores, 1970
4. Petroleum and Organic Chemicals, 1972
5. Natural Organic Materials and Related Synthetic Products, 1972
6. Wood, Paper, Textiles, Plastics and Photographic Materials, 1973
7. Vegetable Food Products and Luxuries, 1975
8. Edible Oils and Fats and Animal Food Products; General Index, 1975

Less detailed, but covering a wider field, is the McGraw-Hill *Encyclopaedia of Science and Technology* (15 vols, 3rd edn, 1971), which is thus especially useful for non-chemical or marginal subjects. This has been supplemented by a number of year-books and a *Dictionary of Science and Technical Terms* (1974). Perhaps more readily available in this country is the older Thorpe's *Dictionary of Applied Chemistry* (12 vols, 4th edn, Longmans, Green, 1937–56). In this case, dictionary is a misnomer, since each subject is treated at far greater length than this suggests. There are a score or more chemical dictionaries proper (e.g. *Concise Etymological Dictionary of Chemistry* (S. C. Bevan, S. J. Gregg and A. Rosseinsky, Applied Science Publishers, 1976) or G. G. Hawley's *Condensed Chemical Dictionary* (9th edn, Van Nostrand Reinhold, 1977)) and a large number of interlingual chemical or scientific dictionaries. (The first chemical dictionary, W. Johnson's *Lexicon Chymicum*, was published in 1692.) Dictionaries and encyclopaedias are very useful sources of physical and analytical data (cf. Chapter 8 of *The Use of Biological Literature* and Chapter 8 of this book). *Kirk–Othmer* is also a useful source of (American) technico-economic data.

One of the classical German works in this area is the six-volume *Römpps Chemie Lexikon* (revised by O. A. Neumüller, 7th edn, Frank'sche Verlag, Stuttgart, 1971–74). This now has a good index in English and ample cross-referencing and is a useful source of information on German (and other) trade names. The *Encyclopaedia of Chemistry* (C. A. Hampel and G. G. Hawley, Eds, 3rd edn, Van Nostrand Reinhold, 1973) contains some 800 articles by 600 authors and is a useful single-volume quick reference source, as is Van Nostrand's *Scientific Encyclopaedia* (5th edn, 1976). Several other more specialized dictionaries and encyclopaedias are discussed in subsequent specific subject chapters. L. M. Myall and D. W. A. Sharp's *A New Dictionary of Chemistry* (4th edn, Longmans, 1968) is useful for its biographical notes.

One of the most useful single-volume compilations is the *Merck Index of Chemicals and Drugs* (9th edn, Merck, Rahway, NJ, 1977), which indicates preparation and properties of some 10 000 chemicals and has an index of formulae, organic name reactions, conversion factors, etc. Where information on the properties or official methods of testing and analysis of a chemical of pharmaceutical importance is sought, this can often be found in current editions of the pharmacopoeias of the various countries or the *British Pharmaceutical Codex*. *Martindale's Extra Pharmacopoeia* (27th edn, Pharmaceutical Press, 1977) is similarly useful, particularly for information on proprietary products. A good guide to sources of natural products is W. Karrer's *Konstitution und Vorkommen der organischen Pflanzenstoffe* (Birkhäuser, 2nd edn, 1976). It has a compound and a plant index but specifically excludes alkaloids. (For further details of works mentioned in this paragraph and similar examples, see Chapter 8 in *The Use of Biological Literature* or *Guide to Drug Information* (W. Sewell, Drug Intelligence Publns, Hamilton, Ill., 1976).) Another type of chemical dictionary to which the experienced, as well as the inexperienced, chemist must sometimes refer is a name index of organic reactions. For examples of books of this type see Chapter 12. That by Krauch and Kunz is probably the most useful for background information. Also useful is D.W.G. Ballantyne and D.R. Lovett's *Dictionary of Named Effects and Laws* (3rd edn, Chapman and Hall, 1970) and the section in the *Merck Index*.

A vast amount of information is provided in the *Colour Index* compiled by the Society of Dyers and Colourists (6 vols, 3rd edn, 1971–). Each dye or pigment is assigned a Colour Index number (according to its chemical type), which is often quoted to avoid ambiguity. In Vols. 1–3 dyes are arranged according to use and information on application, fastness, general data and trade names is given. Vol. 4 gives structural formulae (where available), solubility information and literature references to preparations. Information on intermediates, reducing agents, brighteners, etc., is also given, and there is a trade name index and list of manufacturers in Vol. 5. It is kept updated by quarterly supplements.

MONOGRAPHS

Monographs, being comprehensive surveys of current knowledge on a *single* subject, can be issued more frequently than new editions of large treatises covering a broad field, such as *Beilstein, Gmelin*, etc. They are often issued as members of a series under a general editor or editorial board, each being the work of an individual author. At least 50 such series exist, sponsored by commercial publishers as well as learned societies.

One of the best-known is the *Advances in Chemistry Series*, now published by the American Chemical Society. Many of these are based on papers given at ACS national meetings. The latest addition to the series can be located from advertisements in recent issues of *Journal of the American Chemical Society* or of *Chemistry in Britain*. An even older established series bearing the ACS imprint is published by Van Nostrand Reinhold. Many of these monographs are written by distinguished authors and are so well documented as to be useful keys to the literature.

British monograph series which report symposia proceedings are published by the Chemical Society, Society of Chemical Industry, Biochemical Society, etc. The Royal Institute of Chemistry performs an invaluable service to education through its series of *Monographs for Teachers*, in which elementary topics are treated in a modern and lucid manner.

Monograph series are produced by a number of commercial publishers, including McGraw-Hill, Wiley-Interscience, Van Nostrand Reinhold, Academic Press, etc., in the USA and Pergamon Press, Butterworths, Applied Science Publishers, etc., in Great Britain.

One of the drawbacks to giving lists of monographs, etc., is that they quickly become out of date. This is especially true in rapidly expanding fields such as polymers or biochemistry. One should be cautious about accepting, as being currently valid, opinions expressed or techniques described in a monograph or textbook which is more than five to ten years old. Occasionally a book appears, usually on a theoretical subject (e.g. Lewis and Randall's *Thermodynamics*, 1923), which will influence a generation and then not be out of date. Prior to the post-war instrumental revolution in analytical chemistry, many monographs and compilations were standard works for several decades. A most useful service could be performed by such journals as *Journal of Chemical Education* or *Education in Chemistry* were they to publish (in addition to their excellent book reviews) annotated lists of selected current monographs compiled by one or more experts in the field surveyed. A rapidly growing field such as organometallic chemistry or instrumental analysis should be surveyed every four or five years; less active ones could perhaps be dealt with decennially. The above two journals, *Nature, Chemistry in Britain, Chemistry & Industry, Science Progress, Angewandte Chemie* and many specialist journals publish book reviews which should be perused regularly so that any titles of interest can be noted. These reviews, however, usually appear a long time after the book has been published. They can be located through the monthly *Technical Book Review Index*, which contains abstracts of reviews from British and American journals. An annual index is provided. If the author's name is known, book reviews can also sometimes be located from the *Science Citation Index* (Chapter 5).

Examples of monographs on inorganic, organic and polymer chemistry are given in Chapters 9, 12 and 13, respectively. The titles of recent monographs, etc., in one's field of interest may often be found by looking in the subject catalogue of a large and comprehensive technical library or, for English-language titles, by consulting the *Subject Guide to Books in Print* (Bowker, annually), or other similar guides mentioned in Chapter 2. A particularly useful list (of mainly US books) is published in each September issue of *Journal of Chemical Education*. Bibliographical details of new books are, of course, noted towards the end of the appropriate section in *Chemical Abstracts*. Other general guides are discussed on pages 18—19.

REVIEWS

Reviews appeared in the earliest journals, such as the Royal Society's *Philosophical Transactions*. The first journal devoted solely to reviews appears to have been the *Berlinisches Jahrbuch für die Pharmacie* (1795—1840). There are a number of German review serials which permit us to trace back progress in certain fields of chemistry to 1795, predating and later supplementing the early abstracting journals. The increasing volume of scientific publication has led to a resurgence in demand for periodic comprehensive surveys of a particular field which, although they act as keys to the literature, are tending to complicate information retrieval rather than facilitate it. They now account for about 6% of the literature (Friedman, 1963; Garfield, 1975).

There are a number of sources of review articles such as special review periodicals, review annuals and the ordinary periodicals, especially the news-giving type, e.g. *Chemistry & Industry*. According to a survey of British chemists (Brunning, 1959), the most frequently read review journals were *Annual Reports on the Progress of Chemistry*, *Quarterly Reviews of the Chemical Society* and *Chemical Reviews*. The survey also instanced the desirable criteria of a review as (1) expert writer, (2) critical approach, (3) comprehensiveness, (4) clarity and balance, (5) good bibliography, (6) tabular synopses. Common deficiencies noted were failure to review patent literature and literature in the less known languages, such as Russian, Czech, etc. (The Japanese and the growing Chinese chemical literature also appear often to escape reviewers' attention.) Work published in the Soviet bloc countries is, however, well reviewed by leading Russian chemists in *Uspekhi Khimii* (available in translation since 1960; see Chapter 4). There is also a demand for well-documented surveys of a broader field than is usual in most reviews. For several years VINITI has been publishing such surveys under the title *Itogi Nauki* (*Vistas in Science*), a series which would merit translation.

In addition to those so far mentioned, well-written review articles form an important part of the contents of the following journals: *American Scientist, Coordination Chemistry Reviews, Endeavour, Experientia, Fortschritte der Chemie, Physik und physikalische Chemie, Intrascience Chemistry Reports, Naturwissenschaften, Reviews of Pure and Applied Chemistry* (Australia), *Science Progress, Scientia*, etc. *Angewandte Chemie* publishes authoritative reviews with extensive bibliographies and now has a much wider readership since the English-language edition was introduced (1962). Physics journals such as *Physical Reviews, Reviews of Modern Physics* and *American Journal of Physics* contain a number of reviews of interest to physical chemists. Most of the above journals are general in content but there is an increasing tendency to start up new specialized review journals; a recent example is the quarterly *Progress in Crystal Growth and Assessment*. Reviews of the applications of newer analytical reagents are to be found in the J. T. Baker Co.'s *Chemist-Analyst* (together with many useful laboratory hints). The *Journal of Chemical Education, Education in Chemistry* and the *School Science Review* should not be neglected as a source of reviews by those unconnected with education. The 'technological glossies' such as *Manufacturing Chemist, Food Engineering*, etc., are a useful source of (introductory) reviews.

Special mention must be made of the Chemical Society's *Annual Reports on the Progress of Chemistry*, which started in 1904 and has provided authoritative summaries of the previous year's important chemical papers ever since. Until 1975 the major divisions of chemistry were reported on annually, while perhaps three or four years could go by between reviews of a topic such as thermochemistry. Each section provided one or two hundred references to original papers, including a number presented at symposia and conferences. In order to reduce the length, topics are now reviewed every two or three years, and reviewers are instructed to be more selective in the 'key papers' included. A drawback to using the *Annual Reports* for reconnaissance reading is that quite detailed subject knowledge is still assumed. Provided that one has this background knowledge (obtained, perhaps, from reading appropriate monographs), finding a report on a topic will provide a useful entry to its recent literature, albeit not so comprehensive a one as in earlier volumes and increasingly aimed at the generalist. The *Annual Reports* appear about a year after the end of the year on which they report. A collective index to Vols 1–46 (1904–49) has been issued. From Vol. 64 (covering 1967 literature) *Annual Reports* has been published in two separately available sections after a survey of subscribers claimed to read one-third and admitted ignoring one-third of a volume (Anon., 1967). Section A is on Physical and Inorganic Chemistry and Section B is on Organic Chemistry. A parallel series is published by the Society of Chemical Industry entitled *Reports on the Progress of Applied Chemistry* (1916–).

Chemical Society Reviews has always been oriented towards the chemist wishing to keep up with progress outside his specialist field. This quarterly was formed in 1972 from the merger of the popular *Quarterly Reviews of the Chemical Society* with *RIC Reviews*. In 1969 the Chemical Rubber Co. (CRC Press) introduced a series of quarterly review serials, *CRC Critical Reviews in Analytical Chemistry*, . . . *in Solid State Sciences*, and in several life sciences areas. Major reviews are occasionally updated and published separately as *CRC Monotopics*, e.g. *Laboratory Tests for the Assessment of Nutritional Status* (H. E. Sauberlich, J. D. Skala and R. P. Dowdy, 1974). *Accounts of Chemical Research* (1967–) is an ACS journal containing concise, critical reviews on very narrow topics normally by invited authorities.

The Chemical Society introduced its *Specialist Periodical Reports* in 1967 with *Spectroscopic Properties of Inorganic and Organometallic Compounds* and *Carbohydrate Chemistry*. Over 30 topics are now covered and reviewed on an annual or biennial basis with a reasonably comprehensive literature coverage. During the period 1972–76, Butterworths published a similar series, though using a much higher proportion of overseas editors and contributors. This was called *International Review of Science* (formerly known as the *MTP Series*). This series has now been discontinued.

Often more suitable for reconnaissance reading are articles in a volume having the title *Advances in* . . . or *Progress in* . . . . Books of this type appear more or less annually and 200 titles are noted in *List of Annual Reviews in Science and Technology* (2nd edn, Unesco, 1969). Over 70 chemical titles were listed in the *KWIC Index to Some of the Review Publications, in the English Language, Held at the NLL* (1966), although this contains review periodicals such as *Chemical Reviews*. (The publication of the cumulative index to *Chemical Reviews* with Vol. 60 (1960) has made this source even more useful for older material.)

In general, review serials are of two types: (1) relatively broad topics are reviewed over a short time interval, as in *Annual Reviews of Physical Chemistry*, or (2) a much narrower topic is reviewed over a longer time interval, as in *Advances in Protein Chemistry*. The second type is often the more useful and normally provides a useful literature key as well as occasionally crystallizing a previously nebulous topic. The former are also useful literature keys but can too easily degenerate into a mere bibliography in continuous prose.

Academic Press is the major publisher of those entitled *Advances in*

. . ., while Pergamon Press publish many entitled *Progress in* . . ., most of which are quarterlies.

Advances in Polarography (3 vols, Pergamon, 1959) is the proceedings of the (second) International Congress of Polarography held to honour Heyrovsky's seventieth birthday. It thus comes in the category of *Festschriften*, which, together with *Memorial Volumes*, often contain many useful essays by eminent chemists reviewing the fields in which the scientist thus honoured had worked. *Internationale Bibliographie der Festschriften* (O. Leistner, Osnabruck, 1976) covers 1850 to 1974 and has a key word index.

There are several guides to review literature. *Bibliography of Reviews in Chemistry* (formerly *Bibliography of Chemical Reviews*) (1958–62) ceased publication owing to lack of support. It was recently resurrected as *CARI* (1975–   ) (see Chapter 5). Organic chemistry is particularly rich in reviews and there are two specific guides. The cumulated compilation, originated in ICI by D. A. Lewis, *Index of Reviews in Organic Chemistry* (Chemical Society, 1976), most usefully lists reviews in the form of trade literature in addition to conventional reviews. The other was *Index to Reviews, Symposia Volumes and Monographs in Organic Chemistry* (compiled and edited by N. Kharasch, W. Wolf and E. Harrison, Pergamon Press, 1962) covering the literature from 1940 to 1960. Supplements for 1961–62 (1964) and 1963–64 (1965) were produced. ISI produces an *Index to Scientific Reviews*. (See Chapter 5.)

CONFERENCES, SYMPOSIA AND LECTURES

Attending lectures by eminent chemists can often be a painless way of acquiring background information. Many chemical organizations hold lecture meetings which would benefit far more members than actually attend them. A schedule of such meetings appears in *Chemistry & Industry*, which publishes, as a supplement in January and June each year, a six-months calendar of meetings. Visitors from other cognate societies are, of course, always welcome to attend all meetings other than business ones. Programmes of lectures, short courses, symposia, etc., are organized by the major polytechnics (and increasingly by university extramural departments) and often have several distinguished speakers supported by lesser-known but very competent lecturers. The full text of these lectures is seldom published, though short reports of society-organized lectures may appear in the society's news organ (e.g. *Chemistry & Industry, Chemistry in Britain*, etc.).

The largest chemical meetings are probably the half-yearly ACS meetings, at which several thousand chemists attend a number of parallel sessions. A book of abstracts of the papers presented is published and copies may be obtained in this country. The schedule of papers is published in *Chemical and Engineering News* prior to the meeting. Some of the papers are eventually published in full in the appropriate ACS journal or perhaps in an *Advances in Chemistry Series* monograph (q.v.), but a number never seem to appear again. These abstracts are thus most useful for following current trends in American chemical research. The opposite type of American conference is the Gordon Research Conferences, which are held in the New Hampshire countryside rather than in the large city necessary for the ACS 'conventions'. The Gordon Conferences discourage publication, so that (small numbers of) scientists can meet together to discuss work which is not in a sufficiently advanced state to warrant publication.

Each year sees a growing number of 'International' Conferences. Unfortunately, an increasing number of conference papers are coming into the category known as 'plane ticket' papers (i.e. their main function is to provide their authors with expenses to attend the conference). Plenary lectures and most papers by invited speakers are, however, still worth reading. The full texts of the lectures at these (and occasionally subsequent discussion) are often published afterwards in book form. Those thus published are catalogued by Interdok's *Directory of Published Proceedings* since 1964. The BLLD makes considerable efforts to obtain conference proceedings. It writes to all conference organizers of whom it learns, requesting publication details. As a result a large number of conference proceedings are acquired and these are now listed in the quarterly *Index of Conference Proceedings Received* (1964–   ), which has a KWIC subject index and which cumulates annually. An abbreviated cumulation for 1964–73 is also available. In 1978 ISI introduced the monthly *Index to Scientific and Technical Conference Proceedings*, which cumulates semi-annually and which lists some 80 000 papers from 3000 conferences each year. Unfortunately, the subsequent publication of such lectures is no guarantee that one will be able to locate them through an abstracts service. A survey by Hanson and Janes (1961) has shown that only about half the published papers from a sample selection of conferences were noted in English-language abstracting journals. In addition to those whose proceedings are published in book form (among the leading publishers of such books are Pergamon Press, Butterworths, Plenum Press, Wiley-Interscience, Springer, the several University Presses, etc.), reports and often full texts appear in certain journals. Reports will usually be found in the chemical newspaper(s) of the country in which the conference was held (e.g. *Chemistry & Industry* or *Chemical Age* (UK), *Angewandte Chemie* (Germany), *Experientia* (Switzerland and its neighbours), etc.). In the case of conferences on quite narrow subject areas, one must

search the leading journals and news organs in that field for texts and/
or abstracts of the papers. For example, one would expect to find
reports of papers presented at a conference on synthetic elastomers in
Rubber Age and/or *Plastics* and expect the full papers to be published
in *Polymer Symposia* or *Rubber Chemistry and Technology*, or in *IRI
Transactions* if the Institution of Rubber Industry had organized the
conference.

The dates and venues of forthcoming international conferences can
be found from announcements in news journals or from either Aslib's
quarterly *Forthcoming International Scientific and Technical Conferences* or the Library of Congress's monthly cumulating *World List of
Future International Meetings*, which have subject indexes.

BIBLIOGRAPHIES

Since most new information in science appears in journals, patents, etc.,
rather than books, the term 'bibliography' is used to describe a list of
articles, etc., and/or books. The lists are usually arranged by author, or
subject, or chronologically, and may appear in a journal article or as an
appendix to a book or chapter or may be published in book or pamphlet
form. Bibliographies in book or pamphlet form are the subject of this
section, since the other forms are usually adjuncts to reviews (q.v.).
They are numerous and are published by universities, libraries, and
governmental and industrial organizations. Increasingly bibliographies
are being compiled by manufacturers of research instruments and
chemicals as a sophisticated sales promotion material, e.g. *Sephadex
Literature References*, distributed by Pharmacia AB (Uppsala), who
also issue abstracts on cards.

Two standard bibliographies of bibliographies should be mentioned.
These are T. Besterman's *A World Bibliography of Bibliographies, etc.*
(5 vols, 4th edn, Societas Bibliographica, Geneva, 1965–66), which lists
material to 1963, and the quarterly *Bibliographic Index* (H. W. Wilson
Company), which covers material, mainly in English, published in
books, pamphlets and periodicals from 1937. This cumulates annually
and triennially. A useful reference work for earlier works is C. J. West
and D. D. Berolzheimer's *Bibliography of Bibliographies on Chemistry
and Chemical Technology 1900–1924* (NRC Bulletin No. 50, 1925,
No. 71, 1929, and No. 86, 1932). This and several other bibliographies
are discussed by Burman (1966).

A way of locating readily available bibliographies in book or pamphlet form is to consult the catalogue of a large technical library under
the heading Bibliographies, Special subject area, or under 016:54 or
016:66 where the UDC is in use (see Chapter 2). A number of books,

especially guides to the chemical literature (cf. Burman, 1966), give examples of bibliographies. As discussed in Chapter 5, custom-generated bibliographies are available from the computer-based system at Chemical Abstracts Service, UKCIS or via the on-line services mentioned on pages 85 and 86.

REFERENCES

Anon. (1967). *Chem. in Brit.*, 3, 151
Brunning, D.A. (1959). *Proceedings of the International Conference on Scientific Information*, National Academy of Sciences, Washington, Vol. 1, pp. 545–570
Burman, C.R. (1966). *How to find out in Chemistry*, 2nd edn, Pergamon Press, pp. 35–39
Friedman, H.J. (1963). *J. Chem. Docum.*, 3, 139
Garfield, E. (1975). *Current Comments*, No. 48 (advertising material from ISI)
Hanson, C.W. and Janes, M. (1961). *J. Docum.*, 17, 143

8

The use of standard tables of physical data and other physicochemical literature

R. T. Bottle

INTRODUCTION

Physical chemistry often provides the theoretical basis of our understanding of phenomena associated with either organic or inorganic
compounds or biochemical systems. Although a relatively new science,
it has now virtually reached its adulthood and occupies a position in
the chemical sciences analogous to that achieved by mathematics in
the physical sciences a century ago.

Increasingly it is the basis of investigational techniques used in other
branches of chemistry. Since most books about physicochemical
methods concern systems which fit into the specific subject chapters
of this book or the biochemistry chapter in *The Use of Biological
Literature*, this chapter contains only a small section at the end dealing
with general and practical treatises, selected monographs and texts.
Physical chemistry allows us to quantify relationships and thus the bulk
of the chapter is concerned with compilations of data obtained by
physicochemical techniques and/or required in physicochemical calculations.

A literature search for the best numerical value of a physical property of a substance or material can be a most time-consuming operation. There are many reasons for this, and perhaps the important ones
are, firstly, that many physical properties are determined as a means to
an end (for example, identification) or as a step in the determination of

another property, such as density in a step towards kinematic viscosity determinations. Secondly, physical chemistry is a relatively new science, and in its early years new work — including journals of chemical physics — was published by subscription and so had a limited circulation. Again, many physical properties of materials are ill-defined; the rigidity of a polymer, or the electrical conductivity of an electrolyte, needs careful definition before numerical values can be assigned to it. Thus many details of physical properties are deeply buried in the literature, and effort, patience and time are required to retrieve them. For this reason the standard books of tables are invaluable.

The first edition of Kaye and Laby's *Tables of Physical and Chemical Constants* (1911) was prepared like that of many other compilations by individuals having a strong interest in the properties of matter. Nevertheless the compilation of a book of tables nowadays is beyond the capacity of any one man, and it is now usual to have an editorial board who in turn delegate the collection and scrutiny of data to specialists. These data are afterwards edited and published in a series of volumes. Such a system means that publication occurs when the data are ready. Therefore, it is not possible to follow any order of publication, and this occurs over a period of many years. Often supplementary and additional volumes are published (cf. the fifth edition of Landolt—Börnstein's *Tabellen*).

The problem of compiling data on physical properties, and their critical scrutiny and editing is today a formidable and complex one which can only be solved by collaboration of workers from all over the scientific world and continuous publication year by year. It is in this way that such reference books as the *International Critical Tables*, Landolt—Börnstein's *Zahlenwerte und Funktionen aus Physik, Chemie, Astronomie, Geophysik und Technik*, etc., have been created.

Tables of this type possess three essential characteristics. Each volume, or section, has a concise introduction giving the basic physical facts upon which the data are based. This is followed by the numerical data and finally by a complete series of references to the original literature. They give the critical researcher sufficient information to estimate the accuracy and status of the figures quoted and sufficient references to enable him to set about making a determination for himself if he wishes to do so. All these features are present in any good-quality tables.

MAKING A SEARCH

A search for numerical data from within books of tables can be practised at several levels of competence. The scientist who uses all his scientific training, his experience and imagination, while making a search, will be richly rewarded. The information he will have collected will be much more than the numerical data. He will be aware of physical properties not known, or imperfectly understood, and he may see how to combine the numerical data into a new and useful picture which may offer an explanation that has eluded previous workers and searchers. His knowledge of physical principles and experimental techniques will undoubtedly be extended.

Time is always saved if a little thought is exercised before a search is begun. It is advisable to write down, in detail, exactly what is sought, the definition of the unit in which it is expressed, the order and accuracy required, and the division of physics to which the property belongs. Often a physical property is sought which is not normally expressed directly but which is a function of two or more physical properties that can be given values.

A mathematical training is useful, but a basic knowledge of physics is essential, for those who wish to make a search for the numerical value of physical properties. It is not so much the knowledge of the technique of mathematics as an understanding of mathematical philosophy which enables a search to be conducted quickly and systematically.

When the question has been clearly defined, the next stage is to read the introduction to the appropriate section of the book of tables and if necessary the subject matter in any standard physics text. When an understanding of the subject has been obtained, then a beginning may be made upon standard books of tables.

GENERALIZED CRITICAL TABLES

The International Critical Tables (7 vols and Index, E. W. Washburn, Ed., McGraw-Hill, 1926–33)

At a meeting of the International Union of Pure and Applied Chemistry held in London in 1919 it was decided to prepare these tables under the auspices of the International Research Council and the National Academy of Science, by the National Research Council of the USA under American editorship. They are broadly based upon the International Tables already published annually (see below).

The subject matter is divided into 300 sections, each section being critically examined by the best man available at that time. The tables satisfy all the criteria given above, and the literature up to 1924 was examined in great detail. Nevertheless after 1924 a time-consuming search must be made in *Chemical Abstracts*, etc. (unless one can use Landolt–Börnstein). These tables have been well indexed with full cross-references. This allows easy reference to all subjects and materials and the index is thus the normal method of entering the tables. In a given table chemical compounds are arranged by formula according to a set of key numbers for the elements called the 'Standard Arrangement', which is reminiscent of the Gmelin system (see Chapter 9). (This is explained in Vol. I, page 96, and Vol. III, page viii. In most cases one can find the required compound quite quickly without bothering with the 'Standard Arrangement'.) The bold print numbers in the bibliography following each table are a journal code which will be found at the end of each volume. The tables are quite easy to use and are probably the most suitable source of reference up to 1924.

Landolt–Börnstein's *Zahlenwerte und Funktionen aus Physik, Chemie, Astronomie, Geophysik und Technik* (Springer)

The 6th edition of this internationally known set of tables began publication in 1950. It now appears to be complete apart from the planned general index. A brief synopsis of the contents in English is given as an appendix to this chapter to help those whose German is not fluent select the right volume quickly.

The 5th edition was published in 1923 (2 vols) and brought up to date over the period 1927–36 by supplements (3 vols in 6 parts). This and earlier editions were known as the *Tabellen* and often purchased by individual scientists.

No 7th edition will be published but the data of the 6th edition will be supplemented by a new series of volumes of a narrower subject field, published as the need arises and when the data are accumulated. This is called *Zahlenwerte und Funktionen aus Naturwissenschaften und Technik* and this is under the general editorship of K.-H. Hellwege. Each volume has its preface, table of contents and introductory chapters in both English and German. The six sections and those volumes which have already appeared are listed in the appendix.

The introductory sections of all volumes and parts are comprehensive: indeed, if a basic knowledge of physical principles is assumed, a reader can use the introductory sections as a sound textbook of physics. The literature references are exhaustive and allow the reader to examine the original data upon which the tables are based and to repeat the work if he thinks this should be done. These tables satisfy all the criteria given in the introductory section of this chapter, and are probably the most accurate and up-to-date published data on physical constants.

Annual Tables of Constants (*Tables annuelles de constantes et données numerique de chimie, etc.*)

These tables were begun in 1910 and were published by the International Union of Pure and Applied Chemistry. They were used as a basis, in 1919, for the *International Critical Tables*.

These annual tables are not critical, the object being to collect all numerical data and publish them annually. The first series ran until 1936 in twelve volumes and two index volumes. This series has been continued under the name of *Physico-Chemical Selected Constants* (*New Series*). This series of data has been critically selected by the International Union of Pure and Applied Chemistry and each volume covers a special field with a self-contained bibliography.

The value of these tables is that they are comprehensive. The data of the first series need care in selection (though this was in part done by the editors of *International Critical Tables*), but the new series of Selected Constants are very useful because they are both critical and up to date.

Kaye and Laby's Tables of Physical and Chemical Constants (Longmans, Green)

The 14th edition (1973) of this well-known book of tables has been published under the guidance of an editorial board of physicists drawn mainly from the NPL, who in turn collected critical contributions. SI units have been used since the 13th edition. (SI units are a consistent set based on the m, kg, s and amp set up by the 11th General Conference of Weights and Measures, 1960.)

The data given satisfy all the criteria in the introductory section and cover a wide range of properties in just under 400 pages. The subjects covered are general physics, chemistry, atomic physics and mathematical tables.

Physico-chemical Constants of Pure Organic Compounds (J. Timmermans, Elsevier, Vol. I, 1950, Vol. II, 1965)

This is a critical compilation of all *careful* determinations made on pure organic compounds and has a complete bibliography. Data are collected under compounds, which are arranged according to their composition and functional groups, starting with the paraffins. The easiest way to find a given compound is to use the subject or formula indexes. Volume I covers the literature up to 1950 and Vol. II covers it from 1950 up to 1964 and contains corrigenda to Vol. I.

Physical Properties of Chemical Compounds (*Advances in Chemistry Series*, Nos 15, 22 and 29, American Chemical Society, 1955, 1959 and 1961)

A wide variety of physical properties for nearly 1500 organic compounds are covered in these compilations. No. 29 contains a cumulative index to all three volumes.

HANDBOOKS

This is a well-defined class of tables that contains a wealth of data which have been carefully compiled and edited in an effort to select material to meet the needs of scientific workers who lack the facilities of large technical libraries, which are often not conveniently near manufacturing centres. As a result every effort is made to select the most reliable information and to record and print it with accuracy. In many instances editions are prepared every year and in any case an editorial board is continuously editing, adding data and removing obsolete data from such handbooks. As a result, in the case of handbooks which have passed in a large number of editions, the data offered are accurate and of topical value to all scientific workers.

It is today a practice to publish special handbooks devoted to each well-known industry. To set out to list such a set of handbooks would be self-defeating, but below is given the names and outline of contents of the important general handbooks of data, followed by more specialized ones.

Chemical Engineers' Handbook (5th edn, McGraw-Hill, 1973).

This authoritative reference book covers, comprehensively, the field of chemical engineering as well as important related fields. A considerable amount of data have been taken from the International Critical Tables but these have often been rearranged and recalculated in units used by engineers. The first three editions of *'Perry'* were edited by the late John H. Perry and the 5th edition was edited by R. H. Perry and C. Chilton.

The following sections of physical data are covered:

Units and conversion tables; physical properties of pure substances, specific gravity, melting point, boiling point; vapour pressure of pure substances and solutions; dissociation pressures; densities of pure substances and aqueous inorganic solutions; thermal expansion; Joule–Thomson effect, critical constants, compressibilities, latent heats of pure compounds, specific heats and thermodynamic properties; freezing points and elevation of boiling points; thermodynamic properties and chemical reaction kinetics; flow of fluids in pipes and channels; viscosity data; technology of fluid dynamics; heat transmission by conduction and convection; radiant heat transmission; heat transfer in evaporation; diffusional operations, distillation, solvent extraction and gas absorption; equilibrium relationships; distillation and sublimation; gas absorption and equilibrium data; solvent extraction and dialysis; thermodynamic properties of moist air; electrochemical equivalents.

Handbook of Chemistry and Physics (R. C. Weast, Ed., 59th edn, Chemical Rubber/Blackwells, 1978–79).

Mathematical tables, numerical tables; physical constants of elements, inorganic compounds, organic compounds, alloys, plastics; thermodynamic constants of elements, oxides, hydrocarbons; thermal expansion, vapour pressure, heat conductivity; acoustics; velocity of sound and sound absorption; electrical characteristics; units and conversion factors; miscellaneous basic physical data; sources of data.

The first edition of this well-known and very reasonably priced laboratory companion for chemists and physicists was published in 1914. Recently new editions have been issued biennially and with the 44th edition a considerable enlargement occurred.

Handbook of Chemistry (J. A. Dean, Ed., 11th edn, McGraw-Hill, 1973).

> Physical properties of elements, minerals, organic compounds, industrial materials; miscellaneous tables of specific properties, solubility, density, electrical properties, refractivity, crystal structure, hydrometry, vapour pressure, thermal properties, surface tension, viscosity, compressibility and expansion; numerical tables.

This is known as 'Lange's' *Handbook*, and like the 'Chemical Rubber' *Handbook, International Critical Tables*, '*Landolt–Börnstein*', etc., contains 'inverted' tables where the arrangement is according to the magnitude of a particular physical property and not by substance. Such indexes of melting and boiling points are well known for identifying compounds but indexes of densities, refractive indexes, etc., are also useful. A particularly useful collection of such tables is contained in *Handbook of Tables for Organic Compound Identification* (3rd edn, Chemical Rubber, 1967).

A reasonable amount of data and background information will be found in the *American Institute of Physics Handbook* (D. E. Gray, Ed., 3rd edn, McGraw-Hill, 1972).

ENCYCLOPAEDIAS AND DICTIONARIES (see also Chapter 7)

The newer encyclopaedias and dictionaries often contain physical data which are not available elsewhere, because many of the contributions are written by authorities on their subject who include hitherto unpublished information with the usual criteria of accuracy. Since their arrangement is usually alphabetical, this permits easy and quick reference for physical data.

Encyclopaedic Dictionary of Physics (7 vols and index, J. Thewlis, D. J. Hughes and A. R. Meetham, Eds, Pergamon Press, 1961–63, 5 supplements, 1966, 1967, 1969, 1971 and 1975). The material is arranged alphabetically.

Handbuch der Physik/Encyclopaedia of Physics (S. Flügge, Ed., Springer, 1956–76)

Many of the 54 volumes are in several parts. The *Handbuch* is arranged by topics and contains a considerable amount of data but is perhaps more useful as a source of references to data. Articles are in either German, English or French, with English dominating the later volumes. Some volumes, such as the ones on spectroscopy, could be used as advanced treatises by physical chemists.

In *New Dictionary of Physics* (A. Isaacs and H. J. Gray, Eds, 2nd edn, Longmans, 1975) the data are alphabetically arranged, carefully edited and critically surveyed. Other useful and similar sources are *Van Nostrand's Scientific Encyclopaedia* (5th edn, Van Nostrand Reinhold, 1976) and *The Condensed Chemical Dictionary* (G.G. Hawley, Ed., 9th edn, Van Nostrand Reinhold, 1977). The *Encyclopaedia of Electrochemistry* (C.A. Hampel, Ed., Reinhold, 1964) is an invaluable source of data in its field. Analogous sources of data at the biochemical end of the chemical spectrum are described in Chapter 8 of *The Use of Biological Literature.*

SPECIALIZED COMPILATIONS

A wide range of chemical materials is being produced by industry in ever-increasing quantity and diversity. In order to utilize such materials, data – particularly on physical properties – are collected and often edited and published by the manufacturers, and almost invariably they have information available upon enquiry (see also Chapter 15).

As well as industrial laboratories, government laboratories also produce and collect physicochemical data. Prominent in these activities is the US National Bureau of Standards, whose publication *Journal of Research of the NBS, Section A*, often contains physicochemical data. Up to 1959 the NBS published *Circulars*, two of which are invaluable to physical chemists – *C500, Selected Values of Chemical Thermodynamic Properties* (F. R. Rossini *et al.*, 1952, 1268 pp.) and *C510, Tables of Chemical Kinetics, Homogeneous Reactions* (1951, 731 pp., Supplement 1, 1956, 472 pp.). In *C500* 'best' values are given for 'all inorganic compounds and organic compounds up to 2 carbon atoms' of heats and free energies of formation, entropies, heat capacities, heats and temperatures of fusion, vaporization and sublimation, etc., together with extensive literature references. The material in *C500* was comprehensively revised in *TN270-3* to *TN270-8* (1968–75). An earlier Circular, *C461, Selected Values of Properties of Hydrocarbons* (1947) was revised in 1964 (3 vols).

C510 is a critically evaluated compilation of rates and rate constants and the data are arranged in order of increasing complexity of the key reactant. Many literature references are given and the above compilations satisfy the criteria set out earlier. NBS *Monograph 34* (Vol. 1, 1961, and Vol. 2, 1964) is a further supplement to *C510*. Gevantman and Garvin (1973) have produced a comprehensive listing of compilations of kinetics data for the NBS. Kinetics data are also contained in *Tables of Bimolecular Gas Reactions* (A. F. Trotman-Dickenson and G. S. Milne, NSRDS–NBS 9, 1967). The NBS also published compilations of X-ray data.

In 1963 the NBS set up the National Standard Reference Data System (NSRDS) to integrate the data-collecting and distributing activities of NBS, NASA, AEC, NSF, etc., and to produce a mechanized data bank. A *Property Index to NSRDS Data Compilations, 1964–1972* was compiled for NBS in 1975. The American Institute of Physics and ACS publish the *Journal of Physical and Chemical Reference Data* (1973–) for NBS. A listing of papers appearing in this journal and NSRDS publications can be found in the back of recent editions of the *Handbook of Chemistry and Physics* (page 120).

Recent years have seen a resurgence of data-compiling activities by government and other agencies. More than 50 were listed in *Continuing Numerical Data Projects, A Survey and Analysis* (M. G. Buck, Ed., 2nd edn, National Academy of Sciences, Washington, DC, 1967), of which a new edition would be welcome.

A number use computer-stored data. For example, *CATCH (Computer Analysis of Thermo-CHemical data) Tables*, compiled by J. B. Pedley at the University of Sussex, are routinely updated from computer-stored and -manipulated raw data. The *Physical Properties Data Service (PPDS)* covers more than 350 pure components and calculates properties for multicomponent systems using recommended mixture rules. Further information is available from BP Chemicals International Ltd or the Institution of Chemical Engineers. Other projects in this field are noted from time to time in *Scientific Information Notes*. The February, 1967, issue of *Journal of Chemical Documentation* was a symposium proceedings where several were described.

Some compilations in the *Advances in Chemistry Series* have been mentioned in an earlier section. *Nos 6* and *35* (1952 and 1963) dealt with *Azeotropic Data*, which have been updated and replaced by No. 116 (1973). This includes five-component systems in the 17 000 systems covered and has a formula index. A 1961 compilation by A. W. Francis provides data on *Critical Solution Temperatures (No. 31)* for over 6000 systems. *No. 18* is a compilation of the *Thermodynamic Properties of the Elements* (1956) by D. R. Stull and G. C. Sinke.

Two Russian collections of thermodynamic data are *Termodinam-ichyeskiye Svoistva Individual'nyikh Vyeshchyestvo* (Izdat. Akad. Nauk SSR, 1962) and *Termodinamichyeskiye Konstanti Vyeshchyestvo* (V.P. Glushkov, Ed., 1965–). The former appeared in two volumes, the first of which dealt with theory and calculations of thermodynamic properties and contained 4392 literature references; the second contains tables of thermodynamic properties, mainly of simple molecules. The latter will appear in 10 parts and covers heats of formation, heat capacities, entropies, etc.

Another Russian work (edited by V. V. Kaprov in 1961) has been revised and translated as *Solubilities of Inorganic and Organic Compounds* (H. and T. Stephen, Eds, Pergamon Press, 1963). Volume I (2 parts) deals with binary systems and both aqueous and non-aqueous solvents. Volume II (3 parts) covers multicomponent systems. The classic work in this field is *Seidell's Solubilities*. The 4th edition was completed in 1966 by the American Chemical Society's publication of Vol. II (compounds of elements from K to Z) of *Solubilities of Inorganic and Metal Organic Compounds* (W. F. Linke, Ed.). Volume I (1959) and the earlier editions were published by Van Nostrand. It satisfies the criteria set out in the introductory section of this chapter and covers the literature up to 1956. Solubilities figure high in a list of data most frequently sought by American chemists (Weisman, 1967). They were also high on a list of desirable subjects for comprehensive critical compilations. This has also been studied by M. Slater, A. Osborn and A. Presanis, *Data and the Chemist* (Aslib, 1972). Vapour pressures, densities, viscosities, etc., for a large number of mixtures are contained in J. Timmermans' *Physico-chemical Constants of Binary Systems in Concentrated Solutions* (4 vols, Interscience, 1959–60). The last volume contains the bibliography and formula index for the whole work.

Electrochemical Data (D. Dobos, Elsevier, 1975) is a compilation for laboratory use. Many of the older classics, such as W. M. Lattimer's *Oxidation Potentials* (2nd edn, Prentice-Hall, 1952) or H. S. Harned and B. B. Owen's *Electrolyte Solutions* (3rd edn, Reinhold, 1958), are still useful sources of electrochemical data. The *Atlas of Electrochemical Equilibria in Aqueous Solution* (M. Porbaix, Pergamon Press, 1966), which is a translation of the 1963 French edition, is a unique source of information on such equilibria. Mainly polarographic data are given in *Electrochemical Data* (L. Mietes, P. Zuman *et al.,* Wiley, Vol. A, 1974). The non-aqueous electrometric data literature up to 1973 is distilled into *Nonaqueous Electrolytes Handbook* (2 vols, G. J. Janz and R. P. T. Tomkins, Academic Press, 1972–73). *Buffers for pH and Metal Ion Control* (D. D. Perrin and B. Dempsey, Chapman and Hall, 1974) gathers widely scattered data and includes small computer programs for calculating buffer compositions.

Crystallographers frequently require data compilations. The *International Tables for X-ray Crystallography* were compiled under Dame Kathleen Lonsdale's general editorship and published for the International Union of Crystallography by the Kynoch Press. Volume 1 (1952) concerns symmetry groups, Vol. 2 (1959) deals with mathematical tables and programmes and Vol. 3 (1962) contains miscellaneous physical and chemical data.

Such data compilations are eminently suitable for computerization and the work of the Cambridge Crystallographic Data Unit has recently been described by Kennard *et al.* (1975). *X-ray Diffraction Tables* (J.-H. Fang and F. D. Bloss, Southern Illinois University Press, 1966) is a compact and very convenient laboratory tool. The *Barker Index of Crystals* (6 vols, M. W. Porter and R. C. Spiller, Heffer, 1951–64) is an index of crystal angles which can be used for the classification and identification of crystals. Probably the most comprehensive work in this field is R. W. G. Wyckoff's *Crystal Structures* (6 vols, 2nd edn, Wiley, 1963–71). Short descriptions as well as tables and diagrams are given. The present hardback edition replaces the earlier loose-leaf collection (1948–60).

In certain areas of chemical physics it is possible to program a computer to print-out tables which are required for calculations. An example of such a print-out to aid in the evaluation of Slater orbitals is *Quantum Chemistry Integrals and Tables* (J. Miller, J. M. Gerhauser and F. A. Matsen, University of Texas Press, 1958; Errata, 1959). This is similar to *Tables of Molecular Integrals* (M. Kotani, A. Amemiya, E. Ishiguro and T. Kimura, 2nd edn, Maruzen, Tokyo, 1963).

SPECTROSCOPIC DATA

Spectroscopy affords a good example of both fundamental investigations of phenomena and the application of the techniques as a tool for analytical or structural determinations by organic and inorganic chemists, biochemists, etc. Although most of the data compilations are mainly used for the latter purpose, they are dealt with here since they are largely numerical or graphical in form.

A useful basic introduction (and some data) is provided by R. M. Silverstein and G. C. Bassler's *Spectrometric Identification of Organic Compounds* (3rd edn, Wiley, 1974), which covers infra-red, ultra-violet, nuclear magnetic resonance and mass spectra. The standard text is L. J. Bellamy's *The Infrared Spectra of Complex Molecules* (Vol. 1, 3rd edn, Chapman and Hall, 1975), which contains comprehensive correlation charts. Further details of this and other books dealing with practical and interpretative aspects are given in the physical methods section of Chapter 12.

Compilations of infra-red data ranked fourth (after boiling and melting points and solubilities) in a list of physical data most frequently consulted by American chemists (Weisman, 1967). There are several collections to choose from. H. A. Szymanski and R. E. Erickson's *Infrared Band Handbook* (rev. edn, 2 vols, Plenum Press, 1970) is a hardback one but the restricted number of compounds and rather broad band divisions limit its practical usefulness. More useful (but more expensive) is the Irscot System (*Infrared Structural Correlation Tables*, R. G. J. Miller and H. A. Willis, Heyden, 1966–). This consists of loose-leaf binders containing correlation tables and cards relating the frequency and intensity of infra-red bands to molecular structure. The correlation tables of the Master Index are designed for rapid comparison with an unknown spectrum. Individual bands in the tables are keyed to one or more data cards and one then looks up the individual cards appropriate to the bands of interest and these contain the necessary more detailed information and often literature references. The collection is updated periodically.

Another medium-sized collection in microform is *Infrared Spectra of Selected Chemical Compounds* (R. Mecke and F. Langenbucher, Heyden, 1966). All the 1900 simple organic or inorganic compounds in the Mecke collection are of known structure. There are alphabetical, formula, molecular and serial number indexes and an index of chemical classes. The Irscot and Mecke collections have been critically reviewed by Katritzky (1967a).

The *Aldrich Library of Infrared Spectra*, compiled by C. J. Pouchert (1970), selected 8000 spectra from the files of a leading American laboratory chemicals supplier; Ralph N. Emanuel Ltd is Aldrich's UK agent.

Even larger and more expensive are the *Sadtler Reference Spectra Collections*, published by Sadtler Research Laboratories, Philadelphia, or Heyden, London. The twelve Standard Collections are updated annually and include 51 000 IR, 40 000 UV and 24 000 NMR spectra. There are also less regularly updated Special Interest Collections and Commercial Compounds Spectra covering drugs, polymers, surfactants, etc. All are in loose-leaf binders or in microform. Each spectrum shows the purity and source of the sample and technique used and is identified by its serial number in the several indexes. These are (a) Total Spectra Index, which has four parts (Alphabetical, Serial Number, Molecular Formula and Chemical Class); (b) Infrared Spec-Finder (based on wavelength of strongest band); (c) Ultra Violet Spectra Locator (based on λ_{max} values); and (d) NMR Chemical Shift Index. There are also indexes for each series. Katritzky (1967b) has reviewed the Sadtler Collection of IR spectra.

Documentation of Molecular Spectroscopy (*DMS*) was an Anglo-German venture; the English version published by Butterworths and the German one by Verlag Chemie from 1957 to 1972 containing some 25 000 spectra. Three sets of double edge punched cards were provided: red organic spectra cards, blue inorganic spectra cards and yellow ones for theory, apparatus, etc.; information on both structural features and up to ten strong bands were encoded. *DMS* was, in effect, a self-indexing abstracting system with about 2000 cards per year and contained a number of unpublished spectra. A second set of cards covering NMR (nuclear magnetic resonance), NQR (nuclear quadrupole resonance) and EPR (electron paramagnetic resonance) was published. The publishers of *DMS* have produced a *UV Atlas of Organic Compounds* (1966–67), in five loose-leaf binders each containing about 200 spectra. Spectra are indexed by thirteen basic chromophoric groups as well as by formula and alphabetically.

The Sadtler Collections together with DMS, API/TRC, Coblenz Society, IR Committee Japan and Aldrich (see above) collections and spectra from the literature, a total of 145 000 IR spectra, have been indexed by the American Society for Testing Materials (ASTM) with alphabetical and molecular formula lists. Data base tapes are also available from Heyden, ASTM's European agents. ASTM published *Infrared and Ultraviolet Spectral Index Cards, Empirical Formula-Name Index Cards* and a *Molecular Formula List of Compounds, Names and References to Published Infrared Spectra* (STP 331). The ASTM collections form the basis of what is claimed to be the world's largest generally available spectra collection at the Scientific Documentation Centre, Dunfermline. The SDC now has facilities for searching over 250 000 spectra and 80 000 of these are available for digital searching by spectrum and/or structure. These services are available at moderate cost to any organization becoming a member of the SDC.

Varian Associates' *NMR Spectra Catalog* (2 vols, Palo Alto, Calif., 1962–63) gives spectra for 700 compounds and is provided with name, functional group and shift indexes. Another NMR equipment manufacturer, JEOL, has published *High Resolution NMR Spectra*, which contains 35 ^{19}F spectra. Both Varian and JEOL spectra are included in the Sadtler collections. *Organic Electronic Spectral Data* (Wiley-Interscience) is now an annual publication but Vol. I (1960) covered the years 1946–52. Some 70 journals are scanned for ultra-violet and visible spectra; the wavelengths of all maxima, shoulders and inflections, extinction coefficients and references are given; and the material is arranged by a Hill system formula index. Each volume has so far had a different but expert editor and contains some 25 000 entries. L. Láng has edited a loose-leaf collection of *Absorption Spectra in the Ultraviolet and Visible Region* (20 vols + indexes, Hungarian Academy of Sciences/ Academic Press, 1959–76). Now discontinued, cumulated indexes were issued every five volumes. Another Eastern bloc collection is J. Holubek and O. Štrouf's *Spectral Data and Physical Constants of*

Alkaloids (Heyden). Volume 1 (1965) contains 300 cards and Vol. 2 (1966) contains 100 cards in loose-leaf binders. Melting points, specific rotations and pK values are included in the physical constants. 900 IR and numerous NMR and UV spectra are given in *Atlas of Steroid Spectra* (W. Neudert and H. Röpke, Springer, 1965). It also gives data on melting points, optical rotations and dipole moments.

The *CRC Atlas of Spectral Data and Physical Constants* (J. G. Grasselli and W. M. Ritchley, Eds, 1975) gives data for some 21 000 compounds. Spectra are mainly from the Coblenz or Sadtler collections and other data from the CRC *Handbook* (page 120). Volume I contains a Name/Synonym Directory, Spectra Cross Correlation Tables, etc.; Vols II–IV, data on melting points, densities, refractive indexes, [α], solubility key, IR, UV, NMR, and mass spectra and WLN, etc; Vol. V is an index containing formulae, molecular weights and WLN, etc., though spectra data are indexed in Vol. VI.

About 100 compounds are listed in *Handbook of Fluorescence Spectra of Aromatic Molecules* (I.B. Berlman, Academic Press, 1966). *Mass and Abundance Tables for Use in Mass Spectrometry* (J. H. Beynon and A. E. Williams, Elsevier, 1963) is a computer-compiled aid for identifying ions (containing C, H, N or O only) up to a mass number of 500.

Another type of compilation is the literature index. A number have been compiled by H. M. Hershenson – *Ultraviolet and Visible Absorption Spectra Index*, Vol. I, 1930–54 (Academic Press, 1956), Vol. II, 1955–59 (1961) and Vol. III, 1960–63 (1966); *Nuclear Magnetic Resonance and Electron Spin Resonance Spectra*, 1958–63 (1965); *Infrared Absorption Spectra Index*, Vol. I, 1945–57 (1959), Vol. II, 1958–62 (1964). The Ministry of Aviation Technical Information and Library compiled *An Index of Published Infra-red Spectra* (2 vols, HMSO, 1960). A recent addition is *Mössbauer Effect Data Index Covering the 1975 Literature* (J. G. and V. E. Stevens, Eds, Plenum Press, 1976), which also contains reviews. These are all essentially retrieval guides and are little use without the supporting journals, etc. Even more obvious literature keys to spectroscopic methods and applications are Hilger and Watts's *Spectrochemical Abstracts* (7 vols, 1933–61) and ASTM's *Index to the Literature on Spectrochemical Analysis*, 1920–55. A computer-produced monthly guide to current literature is now compiled by the Mass Spectrometry Data Centre, Aldermaston. This is *Mass Spectrometry Bulletin* (HMSO, 1966–).

GENERAL TREATISES AND TEXTS

Undergraduate texts are numerous and in recent years there has been an increasing tendency to fragment to composite text and issue books confined to certain aspects such as thermodynamics, kinetics, etc. Most of the general texts are American in origin and it should be remembered that they normally use a different set of thermodynamic symbol and electrochemical sign conventions from British and European chemists. One such American text which has been very widely used is W. J. Moore's outstanding *Physical Chemistry* (5th edn, Longmans, 1972), which took over the position formerly held by S. Glasstone's *Textbook of Physical Chemistry* (2nd edn, Macmillan, 1948). Another book which deserves mention because of its unconventional approach (it is in effect an essay on the underlying philosophy of the subject by one of its giants) is the late Sir Cyril Hinshelwood's *Structure of Physical Chemistry* (Oxford University Press, 1957).

There is unfortunately no up-to-date general treatise. J. R. Partington's *An Advanced Treatise on Physical Chemistry* (4 vols, Longmans, 1949–53) contains some still useful data. Academic Press published *Physical Chemistry: An Advanced Treatise* (H. Eyring, D. Henderson and W. Jost, Eds, 1967–75) in 11 volumes but it lacks any volumes on surface chemistry. Physical chemistry is, however, abundantly covered by monographs on a narrow well-defined topic which sometimes form part of a series. The most ambitious of these was Pergamon's *International Encyclopaedia of Physical Chemistry and Chemical Physics*, under the general editorship of E.A. Guggenheim, J.E. Mayer and F.C. Tompkins and with an impressive editorial advisory board. The subject matter was divided into 22 multi-volumed topics, each with its own editor. Publication started in 1960 but halted in 1974. Academic Press have so far produced some three dozen volumes in their series of *Physical Chemistry* monographs. Out-of-date monographs and treatises can to a limited extent be brought up to date through the *Annual Review of Physical Chemistry* (1950–), etc.

The best guide to most practical techniques in physical chemistry is A. Weissberger's *Technique of Chemistry*, especially *Vol. I, Physical Methods*. The three parts of this volume are very well documented and have been through several editions. For further details see Chapter 12. Also still useful for basic techniques is *Physico-chemical Measurements* (3 vols, J. Reilly and W. N. Rae, Van Nostrand, 1948–54).

REFERENCES

Gevantman, L.H. and Garvin, D. (1973). *Int. J. Chem. Kinet.,* 5, 213
Katritzky, A.R. (1967a). *Chem. & Ind.,* 601
Katritzky, A.R. (1967b). *ibid.,* 2140
Kennard, O., Watson, D., Allen, F., Motherwell, W., Town, W. and Rodgers, J. (1975). *Chem. in Brit.,* 11, 213
Weisman, H.M. (1967). *J.Chem. Doc.,* 7, 9

APPENDIX: CONTENTS OF THE 6th EDITION OF LANDOLT–BÖRNSTEIN'S TABLES

Volume 1 – Atomic and molecular physics (5 parts)

Part 1 (1950). Atoms and ions

The system of units: length, mass, time, mechanical, electrical, thermal, photometry; relationship between electrical and magnetic units. The basic constants of physics, velocity of light, Planck's constant, gravitation constant; proton, α-particle. Wavelength of atomic spectra; Faraday effect and other effects due to the external shell of electrons.

Part 2 (1951). Molecular structure

Atomic distances and structure; energy of chemical bindings; dissociation of di- and polyatomic molecules; oscillation and rotation of molecules; the restraint of molecular rotation.

Part 3 (1951). The external electron ring

Band spectra of diatomic and polyatomic molecules; light absorption of solutions in visible and ultraviolet; energy of ionization; optical rotation; electrical moments of molecules; electrical and optical polarization of molecules; magnetic moment; dimagnetic polarization; quantum yield in photochemical reactions.

Part 4 (1955). Crystals

Symmetry, crystal class, space groupings; lattice type, structure and dimensions of crystals; ion and atomic radii; lattice energy; internal variations; electron emission of metals and metalloids; X-ray and electron and high frequency spectra of crystals; lattice distortion and absorption in alkali; halogen compounds.

Part 5 (1952). Atomic nucleus and elementary particles

The energy of the atomic nucleus; hyperfine structure of spectral lines; naturally radioactive atoms; rotation of the nucleus and light quanta; cosmic radiation.

Volume 2 – Properties of matter in the various states of aggregation (10 parts)

Part 1 (1971). Thermal mechanical state

SI units; T, V and P measurement; hardness; natural isotopes; densities and compressibilities of solutions; non-ionized solutions.

Part 2. Equilibria other than melting point equilibria

Part 2 (a) (1960). Vapour condensed phase equilibria and osmotic phenomena

One-component systems: vapour pressures of pure substances; densities of coexisting phases of pure substances; effect of pressure on melting and transition points; liquid crystals. Multicomponent systems: vapour pressures of mixtures; heterogeneous equilibria; colligative properties.

Part 2 (b) (1962). Solution equilibria 1

Multicomponent systems (contd): solution equilibria of (a) gases in liquids, (b) gases in solid and liquid metals, (c) solids and liquids in liquids (solubilities of organic substances in water, etc.; solubilities of inorganic substances in water, etc.).

Part 2 (c) (1964). Solution equilibria 2

Solubilities of organic substances in organic liquids; equilibria in systems with several immiscible phases.

Part 3 (1956). Melting point equilibria and interface phenomena

Melting point diagrams of metal alloys; binary and ternary systems of inorganic compounds; reciprocal salt pairs; silicates; melting point diagrams of organic systems and inorganic—organic systems.

Part 4. Calorimetry (1961)

Molar heat capacities, entropies, enthalpies and free energies and their temperature dependence; tables for calculating thermodynamic functions from known molecular vibrations; Joule—Thomson effect; magnetothermal effect with paramagnetic salts at very low temperatures; thermodynamic functions of mixtures and solutions (including heats of adsorption and neutralization).

Part 5 (a) (1969). Transport phenomena I. Viscosity and Diffusion

Viscosity of gases and liquids; diffusion of gases and liquids.

Part 5 (b) (1968). Transport phenomena II. Kinetics, Homogeneous gas equilibria

Dynamic constants (contd): thermal conductivity; thermal diffusion in gases and liquids; reaction velocities in gases and solids; homogeneous gas equilibria.

Part 6 (1959). Electrical characteristics 1

Electrical conductivity of metals; ionic conductivity in solids (crystals); Hall effect and transistor effect; conductivity transition point; photoconductivity; piezoelectric, elastic, dielectric constants of some systems; dielectric properties of crystals, crystalline solids, glass, synthetic materials (Volume 4, Part 3, 658); crystalline liquids, pure liquids, aqueous solutions; non-aqueous solutions; gases. Thermoelectric effects: Peltier effect; Thomson effect; photoemission and secondary electron emission effects.

Part 7 (1960). Electrical characteristics 2. Conductivity of electron systems

Conductivity of molten salts; pure liquids; conductivity, transport numbers and ion conductivity of aqueous electrolytic solutions; transport numbers; electrolytes in heavy water; conductivity in non-aqueous solutions – organic and inorganic; electrophoretic mobilities and electrokinetic potentials; e.m.f. of reversible and inversible cells in aqueous and non-aqueous solutions; e.m.f. of cells in molten salts; electrolytic dissociation of constants; acid—base indicators; buffer mixtures and compositions.

Part 8 (1962). Optical constants

Metals and alloys; non-metallic substances; glasses and plastics; optical and magneto-optical characteristics of liquid crystals; liquids; gases.

Part 9 (1962). Magnetic characteristics 1

Part 10 (1967). Magnetic characteristics 2

Volume 3 (1952) — Astronomy and geophysics (1 part)

Types and location of instruments; location and time constants; frequency and occurrence of elements; solar system; magnitude of radiation; movement of stars; star systems and special star types; galactic mist. The shape of the earth; force of gravity; minerals and rocks; earth's magnetic field; oceanography; hydrography; meteorology; external atmosphere.

Volume 4 — Basic techniques (4 parts)

Part 1 (1955). Natural materials and mechanical properties

Basic technical and physical units; mass, length, time, area, volume, frequency, velocity; units of force, pressure, energy, work; electrical units; atomic weights; hydrometry and pyknometry. Natural and artificial building materials: cement, mortar, stone, wood, paper, cellulose. Fibrous materials: ceramics, glass, natural and synthetic rubber. Friction and viscosity; static and dynamic friction and viscosity flotation; acoustics.

Part 2. Metallic materials

Part 2 (a) (1963). Fundamentals, Testing methods, Ferrous materials

Characteristics of metals; test methods (investigation of mechanical properties); non-destructive testing; steel production; standards and physical characteristics for ferrous metals; special chemical and physical properties; ferrous and oxygen-free alloys.

Part 2 (b) (1964). Sintering materials. Heavy metals

Powder metallurgy and sintering materials: W, Rh, Ta, Mo, Nb, V, Cr, Co, Ni, Mn, noble metals, Cu, Sb, Zn, Pb, Bi, Sn.

Part 2 (c) (1965). Light metals, Special materials, Semiconductors, Corrosion

Ti, Be, Al, Mg, Li, Rb, Cs; liquid metals, reactor materials, U, Pu, Zr, Hf, Th, rare earths, semiconductors, solders, enamel, hard solders; welding, cutting and extrusion of metals; metal adhesion; corrosion behaviour of materials (arranged by corrosive agent).

Part 3 (1957). Electrical, optical and X-ray techniques

Technical conductivity of solid and liquid materials; thermo-elements; discharge through gases; insulating materials — synthetic and natural; magnetic

---

*These volumes have tables of contents and introductory material in both English and German.

properties; illumination; light sources; luminous substances; light filters; photosensitive materials. X-ray techniques, circuits and dispositions. General and medical X-ray techniques: X-ray spectra; α radiation; fine line structure of X-ray spectra.

Part 4. Determination of thermal and thermodynamic properties

Part 4 (a) (1967). Methods of measurement, Thermodynamic properties of homogeneous materials

Thermometry; hygrometry; thermodynamic properties of gases, vapours, liquids and solids.

Part 4 (b) (1972). Thermodynamic properties of mixtures, Combustion, Heat transfer

Thermodynamic equilibria of mixtures; adsorption. Fuels and combustion. Thermal conductivity.

Part 4 (c). Absorption equilibria of gases in liquids

Part 4 (c1) (1976). Absorption in liquids of low vapour pressure

Water, D_2O, organic liquids, oils, HF, NH_3, etc.; aqueous solutions.

Part 4 (c2) (1977). Absorption in liquids of high vapour pressure

*New Series (K.-H. Hellwege, Ed.)

*These volumes have tables of contents and introductory material in both English and German.

Vol.	3	Luminescence of organic substances (1967)
Vol.	4	Molecular constants from microwave spectroscopy (1967)
Vol.	5	Molecular acoustics (1967)
Vol.	6	Molecular constants from microwave, molecular beam and ESR spectroscopy. Supplement and extension to Vol. II/4 (1974)
Vol.	7	Structure data of free polyatomic molecules (1976)
Vol.	8	Magnetic properties of coordination and organometallic transition metal compounds. Supplement 1 to Vol. II/2 (1976)
Vol. Parts a & b	9	Magnetic properties of free radicals. Supplement to Vol. II/1 (1977) (Part C in preparation)
Vol.	10	Magnetic properties of coordination and organometallic transition metal compounds. Supplement 2 to Vol. II/2

Group III		*Crystal and solid state physics* (see also Group I/Vol. 3)
Vol.	1	Elastic, piezoelectric, piezooptic and electrooptic constants of crystals (1966)
Vol.	2	Elastic, piezoelectric, piezooptic, electrooptic constants, and non-linear dielectric susceptibilities of crystals. Supplement and extension to Vol. III/1 (1969)
Vol.	3	Ferro- and antiferroelectric substances (1969)
Vol. Parts a & b	4	Magnetic and other properties of oxides and related compounds (1970)
Vol. Parts a & b	5	Structure data of organic crystals (1971)
Vol.	6	Structure data of elements and intermetallic phases (1971)
Vol.	7	Crystal structure data of inorganic compounds
Part a		Key elements: F, Cl, Br, I (VII. Main Group)/halides and complex halides (1973)
Part b		Key elements: O, S, Se, Te
b1		Substance numbers b1 ... b1817 (1975)
b2		
Part c		Key elements: N, P, As, Sb, Bi, C
Part d		Key elements: Si, Ge, Sn, Pb, B, Al, Ga, In, Tl, Be
Part e		Key elements: d^9-, d^{10}-, d^1 ... d^3-, f-elements (1976)
Part f		Key elements: d^4 ... d^8-elements (1977)
Part g		References for Vol. III/7 (1974)
Part h		Comprehensive index for Vol. III/7
Vol.	8	Epitaxy data of inorganic and organic crystals (1972)
Vol.	9	Ferro- and antiferroelectric substances. Supplement and extension to Vol. III/3 (1975)
Vol.	10	Structure data of organic crystals. Supplement and extension to Vol. III/5
Vol.	11	Elastic, piezoelectric, pyroelectric, piezooptic, electrooptic constants, and non-linear dielectric susceptibilities of crystals. Revised, updated and extended edition of Vols III/1 and III/2
Vol.	12	Magnetic and other properties of oxides and related compounds. Supplement and extension to Vol. III/4

Group IV		*Macroscopic and technical properties of matter*
Vol.	1	Densities of liquid systems
Part a		Non-aqueous systems and ternary aqueous systems (1974)
Part b		Binary aqueous systems

9

Inorganic chemistry

R. B. Heslop

The aim of this chapter is to guide the reader who wishes to search the literature of inorganic chemistry, including organometallic chemistry, for specific information, and also to draw attention to sources of more general articles and reviews. The emphasis is on publications in English, though a few important sources in other languages are mentioned.

A survey of the field of analytical inorganic chemistry is appended as a separate section of the chapter. The established journals of analytical chemistry do not usually separate the analysis of inorganic compounds from the analysis of organic and organometallic compounds, and for this reason it seems appropriate to deal with analytical aspects in a rather different way at the end. Texts which deal specifically with the analysis of organic compounds are, however, described in Chapter 12.

SEARCHING IN THE LITERATURE FOR SPECIFIC INFORMATION

Anyone wishing to discover information on a specific inorganic compound, or on such a matter as a specialized preparative technique, is well advised to direct his attention first to *Chemical Abstracts*. For information on very recent papers, readers are directed to *Chemical Titles*. Further information on the general topic of searching the literature is given in Chapter 18.

The most valuable treatise devoted to inorganic chemistry is *Gmelins Handbuch der anorganischen Chemie* (*Gmelin*), published by the Gmelin Institute in West Germany. The *Handbuch* began in 1819 with the publication of *Handbuch der theoretischen Chemie* by Leopold

Gmelin. Four further editions appeared in the author's own lifetime up to 1852. In 1921 publication was undertaken by the German Chemical Society and the seventh edition was completed in 1927. The Gmelin Institute, which was founded in 1946, has taken over responsibility for the eighth edition, which, with the addition of supplementary volumes, keeps the literature coverage well up to date.

The system of classification used in Gmelin is based on the sequence of elements given in *Table 9.1*. A system number often includes several volumes with supplements. A compound is located in the volume of

Table 9.1 GMELIN CLASSIFICATION

System number	Element	System number	Element	System number	Element
1	Rare gases	25	Caesium	49	Niobium
2	Hydrogen	26	Beryllium	50	Tantalum
3	Oxygen	27	Magnesium	51	Protactinium
4	Nitrogen	28	Calcium	52	Chromium
5	Fluorine	29	Strontium	53	Molybdenum
6	Chlorine	30	Barium	54	Tungsten
7	Bromine	31	Radium	55	Uranium
8	Iodine	32	Zinc	56	Manganese
9	Sulphur	33	Cadmium	57	Nickel
10	Selenium	34	Mercury	58	Cobalt
11	Tellurium	35	Aluminium	59	Iron
12	Polonium	36	Gallium	60	Copper
13	Boron	37	Indium	61	Silver
14	Carbon	38	Thallium	62	Gold
15	Silicon	39	Lanthanides	63	Ruthenium
16	Phosphorus	40	Actinium	64	Rhodium
17	Arsenic	41	Titanium	65	Palladium
18	Antimony	42	Zirconium	66	Osmium
19	Bismuth	43	Hafnium	67	Iridium
20	Lithium	44	Thorium	68	Platinum
21	Sodium	45	Germanium	69	Technetium
22	Potassium	46	Tin	70	Rhenium
23	Ammonium	47	Lead	71	Transuranic
24	Rubidium	48	Vanadium		elements

the constituent element having the highest system number. Thus barium chloride appears under system number 30, whereas hydrogen chloride appears under system number 6. A twelve-volume *Formula Index* for material in volumes appearing up to 1974 is being published.

Gmelin is a compilation of all the inorganic chemical literature which is available. It is simultaneously a bibliographical compendium and a textbook which attempts to evaluate retrospectively the entire subject matter from the standpoint of modern knowledge and to extract from obsolete material what is of value today. In recent volumes and supplements, headings, sub-headings and tables of contents in English are provided.

The year 1973 saw the publication of two comprehensive treatises in English: *Comprehensive Inorganic Chemistry*, executive editor A.F. Trotman-Dickenson, produced in five volumes by Pergamon; and *Inorganic Chemistry Series One*, consultant editor H.J. Eméleus, produced in ten volumes as part of the *MTP International Review of Science* by Butterworths. The older Mellor's *Comprehensive Treatise on Inorganic and Theoretical Chemistry*, originally published in sixteen volumes by Longmans, Green between 1922 and 1937, is being brought up to date by the issue of supplementary volumes, but the progress made in recent years has been slow, though three volumes on boron appeared in 1979.

As in other areas, comprehensive treatises seem to be giving ground to specialized monographs such as *Boron and Refractory Borides* (V.I. Matkovich, Ed., Springer, 1977).

JOURNALS

Some of the more important general journals published in English in which papers in inorganic chemistry appear regularly are listed below.

Journal of the Chemical Society now appears in five sections, of which one, the
  Dalton Transactions, contains papers in inorganic chemistry.
Journal of the American Chemical Society
Canadian Journal of Chemistry
Australian Journal of Chemistry

There are often communications on inorganic chemistry in *Nature* and in *Chemistry & Industry*. Some authors use these two journals, and also to some extent the *Journal of the American Chemical Society*, to give preliminary notice of work which they intend to publish in greater detail later.

The need for publications which are able to ensure the quick appearance of communications of outstanding interest has been met in the last decade by the *Proceedings of the Chemical Society* (London), which filled this role from 1957 to 1964, and by *Chemical Communications* (part of *Journal of the Chemical Society*), also published by the Chemical Society, since 1965. A more recent entry into this field which is of particular interest to the inorganic chemist is *Inorganic and Nuclear Chemistry Letters* (Pergamon), which began publication in 1966.

Of journals specializing in inorganic chemistry, the oldest is the German *Zeitschrift für anorganische und allgemeine Chemie*, which first appeared in 1892. It now contains tables of contents in English and also publishes some papers in English. The journal *Inorganic Chemistry* has been published since 1962 by the American Chemical Society, and has attracted a rapidly increasing number of papers. Another journal which has expanded quickly since its appearance in 1955 is the *Journal of Inorganic and Nuclear Chemistry* (Pergamon). Mention must also be made of *Inorganica Chimica Acta* (Elsevier). To keep English-speaking chemists abreast of the work of inorganic chemists in the Soviet Union, the Chemical Society, with the support of the Department of Education and Science, produces the *Russian Journal of Inorganic Chemistry* under the editorship of C.C. Addison.

Of the more specialized journals which have appeared in recent years may be mentioned the *Journal of Organometallic Chemistry* and the *Journal of Less Common Metals*, both published by Elsevier, and the *Journal of Coordination Chemistry* (Gordon and Breach).

PERIODICAL REPORTS AND REVIEWS

Since 1967 the *Annual Reports* of the Chemical Society, London, have been divided into two sections, of which Section A contains up to 300 pages on inorganic chemistry in such subsections as Typical Elements, Transition Elements, Carbonyls and Organometallic Compounds, and Mechanisms of Inorganic Reactions. More specialized reviews are also being produced by the society, under the title *Specialist Periodical Reports*, either biennially or triennially. Those of particular interest to inorganic chemists are *Inorganic Chemistry of the Main Group Elements, Inorganic Chemistry of the Transition Elements, Organometallic Chemistry, Inorganic Reaction Mechanisms, Electronic Structure and Magnetism of Inorganic Compounds* and *Spectroscopic Properties of Inorganic and Organometallic Compounds.*

Two publications which have appeared annually since 1959 have kept up a high standard of reporting on research in inorganic chemistry: *Advances in Inorganic Chemistry and Radiochemistry* (Academic Press), edited throughout the period by H. J. Emeléus and A. G. Sharpe; and *Progress in Inorganic Chemistry* (Interscience), edited for many years by F. A. Cotton but more recently by others. These two are similar in format; a volume contains about six review articles each of about 60 pages. The series *Advances in Organometallic Chemistry* (Academic Press) has been edited by F. G. A. Stone and R. West since 1964.

Reviews, in English, which contain occasional contributions on inorganic chemistry and organometallic chemistry are *Chemical Society Reviews, Chemical Reviews, Angewandte Chemie: International Edition* and *Russian Chemical Reviews*, all of which are mentioned elsewhere in this book. Publications of the review type which specialize in topics of interest to inorganic chemists are *Coordination Chemistry Reviews*

(Elsevier), *Organometallic Chemistry Reviews* (Elsevier), *Structure and Bonding* (Springer) and *Transition Metal Chemistry* (Academic/Verlag Chemie).

HANDBOOKS ON PRACTICAL ASPECTS

Practical aspects of inorganic synthesis are well covered in *Inorganic Syntheses* (McGraw-Hill), of which fifteen volumes have appeared since 1939, in *Organometallic Syntheses* (Academic Press), a series which began in 1965; and in *Preparative Inorganic Reactions* (Interscience), which began in 1964. More general aspects of practical techniques are treated in the six volumes of *Technique of Inorganic Chemistry* (Interscience), under the general editorship of H. B. Jonassen and A. Weissburger. Another useful source is *Handbook of Preparative Inorganic Chemistry* (2 vols, 2nd edn, Academic Press, 1965), edited by G. Brauer. It contains a short section on methods as well as about 2400 descriptions of preparations. The physical methods used in determining the structures of inorganic compounds and organometallic compounds are often similar to those used for organic compounds, and appropriate texts will be found in Chapter 12.

TEXTBOOKS AND OTHER CONCISE SOURCES

A number of textbooks which are intended for undergraduate and immediate postgraduate studies contain useful bibliographies. Outstanding in this respect is *Advanced Inorganic Chemistry* (F. A. Cotton and G. Wilkinson, 3rd edn, Interscience, 1972), which contains references to original papers as well as to review articles. *Inorganic Chemistry* (2 vols, C.S.G. Phillips and P.J.P. Williams, Oxford U.P., 1965) is written at a similar level and contains a wealth of information on the physicochemical properties of inorganic compounds which is of particular value to teachers preparing lectures for undergraduates. The newer *Inorganic Chemistry* (R. B. Heslop and K. Jones, Elsevier, 1976) has no pretension to be more than an undergraduate text, but it gives many useful references to review articles and is the only textbook of inorganic chemistry yet available which uses the international system of units correctly and consistently.

Information on structural aspects of inorganic chemistry is presented comprehensively in the excellent *Structural Inorganic Chemistry* of A. F. Wells (4th edn, Oxford University Press, 1975). Its index is a particularly useful source of information on the structures of inorganic molecules and crystals. Mention must also be made of two *Special*

Publications of the Chemical Society: *SP11*, on *Interatomic Distances* (1958) and *SP18*, a 1965 supplement to *SP11* on *Interatomic Distances*.

Two volumes of *Stability Constants* have been published by the Chemical Society. The first volume, edited by L. G. Sillen (for inorganic ligands) and A. E. Martell (for organic ligands), was published in 1964 and covered the literature up to 1962. The second volume, edited by L. G. Sillen and E. Hogfeldt (for inorganic ligands) and A. E. Martell and R. M. Smith (for organic ligands), appeared in 1971 and covered the literature up to 1969. More recently Martell and Smith have begun the production of *Critical Stability Constants* (Plenum Press), the first volume being dated 1974.

The authoritative work on the names of inorganic compounds is *Nomenclature of Inorganic Chemistry* (2nd edn, Pergamon for IUPAC, 1971). Guidance in the use of the rules is given in *How to Name an Inorganic Compound* (W. C. Fernelius, Ed., Pergamon, 1977).

ANALYTICAL CHEMISTRY

General

In 1954 the Society for Analytical Chemistry began publication of *Analytical Abstracts*, which is organized in a very similar way to *Chemical Abstracts*, with an author index and a subject index to each volume. It appears monthly and Section B of each issue deals with inorganic chemistry. Since 1975 its publication has been undertaken by the Chemical Society.

A series entitled *Advances in Analytical Chemistry and Instrumentation* (Interscience) began in 1960 and eleven volumes had appeared up to 1971 under a number of editors.

Journals

Of the journals on analytical chemistry published in English, the two oldest are *The Analyst*, formerly published by the Society for Analytical Chemistry but now the analytical journal of the Chemical Society, and *Analytical Chemistry*, published in Washington, DC, by the American Chemical Society. The latter appeared before 1948 as the analytical edition of *Industrial and Engineering Chemistry*. Both of these journals contain some review articles as well as original communications. Each year the April issue of *Analytical Chemistry* is devoted to reviews of the past year's progress. Two other useful journals which are mainly in English are *Analytica Chimica Acta* (Elsevier) and *Talanta* (Pergamon).

The latter often contains valuable reviews but the former is devoted entirely to original work. *Analytical Letters* commenced in 1968 as a monthly quick publication medium.

In recent years some of the specialized aspects of analysis have had journals devoted to them. Examples of such publications are *Journal of Chromatography* (Elsevier), *Microchemical Journal* (Interscience) and *Journal of Electroanalytical Chemistry and Interfacial Electrochemistry* (Elsevier).

Treatises

In the 1st edition (1959–76) of *Treatise on Analytical Chemistry* (I. M. Kolthoff and P. I. Elving, Eds, 2nd edn, Wiley, 1978–) Part I consists of six volumes on the theory and practice of analytical chemistry, Part II of nine volumes on the analytical chemistry of the elements and four volumes on organic analysis. Part III deals with analytical chemistry in industry.

Comprehensive Analytical Chemistry (C. L. and D. Wilson, Eds, Elsevier) consists of three volumes on classical methods (IA, 1959; IB, 1960; IC, 1962), and three on electrical and physical separation methods (IIA, 1964; IIB, 1968; and IIC, 1971).

The much older work *Standard Methods of Chemical Analysis* (Van Nostrand) was completely revised under the editorship of F. J. Welcher in the period 1962–66 (6th edn). Volume I deals with the elements, Vol. 2 with industrial and natural products and with non-instrumental methods, and Vol. 3 with instrumental methods. Another three-volume work is *Chemical Analysis* (C. R. N. Strouts, H. N. Wilson and R. T. Parry-Jones, Eds, Oxford U.P., 1962). On physical methods there is *Physical Methods in Chemical Analysis*, by W. G. Berl (Academic), of which four volumes appeared up to 1961.

Textbooks

Shorter works which are well stocked with useful information are *Instrumental Methods of Chemical Analysis*, by G. W. Ewing (4th edn, McGraw-Hill, 1975); *Chemical Instrumentation*, by H. A. Strobel (2nd edn, Addison-Wesley, 1973); *Modern Analytical Chemistry*, by W. F. Pickering (Marcel Dekker, 1971); *Vogel's Textbook of Quantitative Inorganic Analysis*, by J. Bassett *et al.* (4th edn, Longman, 1978); and *Vogel's Textbook of Macro and Semimicro Qualitative Inorganic Analysis*, by G. Svehla (5th edn, Longman, 1979).

10

Nuclear chemistry

K. Jones

Since the appearance of earlier editions of this book, it would be broadly true to say that nuclear chemistry has come of age. This has been accompanied, if not necessarily indicated, by a decline in the proportion of innovative papers while applications utilizing available techniques have expanded rapidly together with an increase in the quality and quantity of reviews and books.

Perhaps the first thing that an author is required to do in the field of nuclear chemistry is to define the readership. First, there is the specialist nuclear chemist, who no doubt with experience will have established a reading list and will be able to select what is useful for his particular needs. The second main group comprises chemists of other specializations who from time to time have cause to consult the nuclear chemistry literature, usually with the hope that in due course it will provide a way to solve a problem. However, as well as the trained specialists of any subject, one must also consider the layman, and the social, political and environmental aspects of the subject for and about which much has been written; all this has given the literature a somewhat unique breadth to organize.

It is still true to say that developments in the field of atomic energy have had widespread repercussions in many branches of chemistry (*Chemical Aspects of Nuclear Reactors*, J. K. Dawson and R. G. Sowden, Butterworths, 1963). The nuclear energy industry involves chemistry at many stages. Likewise, radioisotopes require chemists at each stage of their production, assay and application. Thus the range of studies for the nuclear chemist is very wide. The various branches of the subject are considered under the following headings:

(1) Radiochemistry – the application of radioisotopes to problems of research and analysis; the radioisotope is employed primarily as a tracer or marker in the investigation.

(2) Nuclear chemistry – the elucidation of nuclear reactions; the production and characterization of new radionuclides formed in these reactions.

(3) Radiation chemistry – the study of the behaviour of a chemical system exposed to a source of ionizing radiation or fission fragments; this section may also include the chemical effects consequent upon the radioactive decay of a component of the system which result from the recoil of the disintegrating nucleus.

The principal uses of radioisotopes have occurred in radiochemistry and this chapter is mainly a guide to the literature concerned with (a) the selection of the isotope, (b) the provision of laboratory facilities and measuring equipment, and (c) the basic techniques and types of application. Although radioisotopes have been widely used in chemistry, their full exploitation has been to some extent limited by undue emphasis on the radiation hazards and on the expense of laboratory facilities and counting equipment. Careful perusal of the relevant literature should convince any potential user that neither of these two factors will be the overriding one, especially when tracer work in the laboratory is involved; however, rather more specialized facilities will be required for work in nuclear and radiation chemistry.

At the outset, however, it is perhaps important to state the somewhat special position in which workers in the field of nuclear science find themselves. Unlike many other specialized areas of endeavour, radiochemistry and more generally work with any form of ionizing radiation (as defined in the Radioactive Substances Act) is subject to quite specific regulations. This legislation varies from one country to another, but to take the UK as example (further details will be given on page 156), the regulations require notification and registration with the Department of the Environment (schools must similarly notify and register with the Department of Education and Science) of the use and disuse of ionizing radiations, and that one or more 'competent persons' must be appointed to supervise the work and assist in enforcing the Regulations (principally in the keeping of records). Such 'competent persons' will of necessity have undergone special training which will probably have included some guidance in the relevant literature.

It is hoped that this chapter will be of further use to such 'competent persons', particularly in areas outside their immediate experience, but it must be emphasized to other readers that, in contrast to the usual situation with respect to the open chemical literature, knowledge does not give the reader the immediate right to practice.

General sources of information covering all aspects of work with radioactivity and radiation are first of all surveyed; then follow a survey of the literature devoted to radiochemistry, brief surveys with respect to nuclear and radiation chemistry, and finally a guide to the hazards and the principles of radiological protection together with a summary of the legal requirements.

GENERAL LITERATURE

Although a considerable number of publications are devoted almost exclusively to nuclear science, only sources of information of some interest to chemists are referred to in the following pages.

Abstracts

The chemists' normal first choice source of abstracts, *Chemical Abstracts*, only provides limited assistance with nuclear information. Two main sub-divisions are appropriate:

Biochemistry: Section 8 'Radiation Biochemistry'
Physical & Analytical: Section 70 'Nuclear Phenomena'; Section 71 'Nuclear Technology'; Section 74 'Radiation Chemistry'

Sections 70 and 71 are available in the computer-readable file 'Energy' and UKCIS will provide extracts from *CA Condensates* through computer scanning as a *Macroprofile.*

Nuclear Science Abstracts provided up to July 1976 a world-wide coverage of the literature in the whole field of nuclear science and engineering. The abstracts covered the relevant journals (about 800) in addition to the unclassified and declassified documents of the various Authorities for Atomic Energy, the reports from Industrial and Research Associations, Conferences, and the Patent literature. Publication was semi-monthly from 1947. Annual, quarterly, half-yearly and cumulative indexes covering subject, author and radionuclide were also issued. The contents are in systematic subject groups but their principal use now will be for retrospective searching.

As from July 1976, *INIS Atomindex*, published by IAEA in Vienna since 1969, remains the sole comprehensive abstracting service to provide coverage of the world's nuclear literature. Issues appear fortnightly and are supported by Cumulative, Personal Author, Corporate Entry, Subject, Conference, Report Standard and Patent Number Indexes. Member States of the International Atomic Energy Agency have a Liaison Officer through which IAEA publications can be obtained.

The problems of undertaking a literature search involving a considerable number of publications are similar to those in other branches of science; the procedure for undertaking a literature survey is as described for *Chemical Abstracts* (Chapter 5).

Journals

Original papers on nuclear chemical topics are published in a number of specialist journals, and some of these are listed in *Table 10.1*. A comprehensive *List of Periodicals in the Field of Nuclear Energy* is published by IAEA.

Table 10.1 SOME SPECIALIST JOURNALS IN NUCLEAR CHEMISTRY

Title of Journal	Date of first publication	Publisher
International Journal of Applied Radiation and Isotopes	1957	Pergamon Press
Journal of Inorganic and Nuclear Chemistry	1955	Pergamon Press
Radiochemistry (trans. of *Radiokhimiya*)	1960	Pergamon Press
Radiochimica Acta	1962	Academic Press
Physical Reviews	1893	American Physical Society
Physical Review Letters	1958	American Physical Society
Journal of Radioanalytical Chemistry	1968	Elsevier
Radiochemical and Radioanalytical Letters	1968	Elsevier

Unfortunately, there is a growing number of 'deceased' publications; *Isotopes and Radiation Technology* was an excellent source of review articles between 1964 and 1972, but was forced to cease publication through lack of funding.

A general problem in nuclear as in all branches of science is the existence of a large and important body of technical literature in foreign languages. Translations of this literature are undertaken by both government and private agencies, and a useful guide to these services in the US has been published in *Nucleonics* (1961). *Transatom Bulletin* is published monthly by Euratom and lists translations of the nuclear literature prepared by its member organizations. *Euratom Information* is a polyglot publication giving regular information on the scientific and technical publications of the research establishments of the European Atomic Energy Community. UKAEA translations are listed in the monthly *UKAEA List of Publications Available to the Public* (see also Chapter 15).

Articles involving the applications of radioisotopes and radiation in chemical research also appear regularly in the established chemical journals. Articles of a more general character appeared from time to time in the American journal *Nucleonics*, which was published monthly from 1947 to 1967 by McGraw-Hill. A feature of this journal was the annual publication from 1960 to 1967 of a *Reference Data Manual* giving valuable compilations of information and procedures in applied radiation.

Research papers from official sources

The policy of the UK Atomic Energy Authority and other national authorities is to publish research papers as far as possible in the recognized journals. However, many reports and documents are published internally for two reasons: (1) the necessity for rapid internal communication of information, and (2) to minimize the pressure on available space in the journals in view of the frequently specialized nature of the information. Some reports, however, do appear later in the technical and scientific press. The reports from the various national authorities were formerly lodged in a number of depository libraries in the UK as listed in *Table 10.2*. Lists of other depository libraries may be found in *Nuclear Science Abstracts*.

Table 10.2 DEPOSITORY LIBRARIES IN THE UK FOR OFFICIAL NUCLEAR ENERGY REPORTS

British Library Lending Division	Boston Spa
British Library Science Reference Library	London
Science Museum Library	London
Central Library	Birmingham
Central Library	Kingston upon Hull
Central Library	Liverpool
Central Library	Manchester
Central Library	Newcastle upon Tyne
Central Library	Sheffield
Public Library	Nottingham
Ealing Commercial and Technical Reference Library	London (Acton)

Documents may now be seen by scientists, etc., at UKAEA Public Information Centre at Charles II St, London, and copies for public use are available at the Science Reference Library and can be obtained by libraries in the UK as a whole through the British Library Lending Division at Boston Spa.

The location of a report may be found by reference to periodical *Cumulative Lists* published by UKAEA or through *Nuclear Science Abstracts* or *INIS Atomindex*. In the case of USAEC reports, the *Bibliography of Selected AEC Reports of Interest to Industry* is of assistance, available from addresses given by E. J. Brunenkant and R. M. Berg, *Nucleonics*, **19**, (5), 109 (1961). Atomic Energy reports post-1964 are also stored in NTIS data base.

UKAEA reports and documents are now lodged in the Copyright Libraries: the Bodleian Library, Oxford; Cambridge University Library; the National Library of Scotland, Edinburgh; and the National Library of Wales, Aberystwyth. In many cases they can by purchased from HMSO at Government Bookshops or through certain booksellers in the UK. A complete list of UK Atomic Energy publications sold by HMSO is available and a general description of the UKAEA report series with information on documentation and referencing is given in *Guide to UKAEA Documents* (5th edn), edited by J. Roland Smith (1973). A *List of Available Publications* is issued monthly by AERE Harwell; annual cumulations have also been issued since 1956. *Harwell Information Bulletin* is issued weekly and is a list of titles from recent journals and publications received by the Production and other groups of UKAEA.

Review literature

Reviews or monographs provide an excellent starting point to familiarize a reader approaching a new area of chemistry with a pre-digested appraisal. They also constitute perhaps the greatest growth area in chemical literature.

Comprehensive Inorganic Chemistry (executive editor, A. F. Trotman-Dickenson, Pergamon Press, 1973) is a multi-author reference work and literature guide in five volumes offering a systematic and comprehensive source of inorganic data, including radioisotopes and the radioactive elements (see Chapter 9).

Included also under this heading are the Proceedings of various Symposia and International Conferences. Plenary lectures are usually given verbatim together with abstracts of the other contributions. These are far too numerous to list here, and *Publications in the Nuclear Sciences* and updating appendices should be consulted (available from IAEA or HMSO).

Progress in Nuclear Energy (Pergamon Press, 1956–) is an International Review Series involving serial publication of volumes in a number of divisions and incorporating edited proceedings of the two UN Conferences.

Annual Reviews of Nuclear Science (Annual Reviews Inc., 1952–), although biased towards nuclear physics, provides articles of value to the chemist from time to time.

The *MTP International Review of Science* has appeared in two series, of which the Inorganic Series One Volume 8 Radiochemistry (volume editor, A. G. Maddock) was published in 1972 and Series Two in 1975. Unfortunately, it is doubtful whether this Series will continue. During the same period the Chemical Society have had a similar aim with their *Specialist Periodical Reports* (*SPR*). The three volumes on Radiochemistry published in 1972, 1975 and 1976 (G. W. A. Newton, Ed.) provide excellent entries for the topics covered.

Source Material for Radiochemistry NP-19584 (National Academy of Sciences, Washington, DC, 1971) contains over 100 pages of references to books, reviews and journals on a wide range of subjects. *Sources of Information on Atomic Energy* (L. J. Anthony, Pergamon Press, 1966) is an earlier general guide to the literature in this field.

Perhaps the most helpful source of titles is the biannual review on Nucleonics in *Analytical Chemistry*. This review first appeared in 1962, when it was truly comprehensive, and it has evolved over the years through various degrees of justified selection. The coverage is not restricted to analytical applications. (See *Anal. Chem.*, **42**, 123R, 1970; **44**, 438R, 1972; **46**, 431R, 1974; **48**, 96R, 1976.)

RADIOCHEMICAL APPLICATIONS

In reviewing the available literature, it has been presumed that an investigation employing radioisotopes as tracers is to be undertaken for the first time. The feasibility of the operation will require attention to the following matters: (a) selection of the radioisotope, (b) provision of laboratory facilities and counting equipment, and (c) selection of the appropriate isotope techniques. The literature is briefly surveyed under these headings, and then selected references to actual applications in research and chemical analysis are listed.

Radioisotopes

Radioisotopes of the great majority of elements are now available, in addition to a wide range of compounds incorporating radiocarbon, radiophosphorus, radiosulphur, radioiodine and radiotritium. The properties of radioisotopes of importance in selection are:

(1) Half-life – if possible greater than ten days, but not too long for reasons of safety and disposal.

(2) Type and energy of radiation — beta-emitting isotopes of maximum energy greater than 0.5 MeV are ideal for the simplest and most straightforward application.

The availability of a radioisotope is dependent on its method of production. Isotopes which can be produced by neutron irradiation of a stable target element (depending on the neutron capture cross-section of the target) are generally readily available, as also are certain fission products.

The relevant nuclear data can be found in a number of publications:

Table of Isotopes (C. M. Lederer, J. M. Hollander and I. Perlman, 6th edn, Wiley, 1967)

The Radiochemical Manual (2nd edn, The Radiochemical Centre, Amersham, 1966) (NOTE: This manual also contains useful information about the production of radioisotopes and about the special characteristics of radioactive materials such as purity and problems of usage. The Centre also publishes a series of short monographs on radiochemical topics.)

Radioactive Nuclei Decay Schemes (Dzhelepov and Peker, translation editor, D. L. Allan, Pergamon Press, 1961)

Gamma-rays of Radionuclides in Order of Increasing Energy (D. N. Slater, Butterworths, 1962)

Nuclear Data in Science and Technology (IAEA, 1973)

Information regarding commercial availability in the UK is obtainable from:

Radio-active Chemicals: catalogue issued by The Radiochemical Centre, Amersham (gives information about primary isotopes and compounds of common radionuclides)

Standardized Radioactive Solutions 1974/5: catalogue issued by The Radiochemical Centre, Amersham (also includes updated tabulations of properties of various nuclides)

International Directory of Radio-isotopes (2 vols, IAEA, STI/PUB/5 & 7)

Individual elements are surveyed in the series *Radiochemistry of the Elements* (US National Academy of Sciences Monograph Series, NAS-NS-300, etc.) and also as individual texts as cited in the section on review literature.

Useful charts of the radionuclides are as follows:

Trilinear Chart of the Nuclides (W. H. Sullivan, USAEC, Oak Ridge National Laboratory)

Chart of the Nuclides (Federal Ministry of Nuclear Energy, Federal Republic of Germany; published by Geisbach und Sohn Verlag, München 34, Barer Strasse 32)

Laboratory facilities

For operations on a limited scale, part of an existing laboratory can be adapted for the purpose; on a more extended scale, special provision will be necessary. Some guidance in these matters is provided by a number of textbooks listed in the section 'Basic chemical and handling techniques'. A useful guide to the 'Design of radionuclide laboratories' is available (D. Hughes, *Chem. in Brit.*, 4, (2), 63, 1968). It should be noted that at the time of notification of the intention to work with radioactive materials, approval of the facilities may be required through the Factory Inspectorate.

Counting equipment

The Geiger–Müller assembly remains the simplest and cheapest form of equipment where beta-emitting isotopes of sufficient energy are available; the counting efficiency is very low for gamma-radiation and a scintillation assembly is then recommended. The scintillation counter has advantages for low-energy beta emitters and is also an integral part of the gamma-spectrometer, which has special application to activation analysis (page 152). In recent years a considerable amount of more sophisticated hardware has become available for the specialist, but a guide to the selection and operation of equipment may be obtained from the following references:

Radioactivity and Radiation Detection, (D. G. Miller, Ed., Gordon and Breach, 1972)
Nuclear Radiation Detection (W. J. Price, 2nd edn, McGraw-Hill, 1964)
'Measurement techniques and instrumentation' (J. D. Hemingway, Chapter 3 in *MTP International Review of Science*, Series Two, Vol. 8, 1975)
Nucleonic Instrumentation (C. C. H. Washtell and S. G. Hewitt, Newnes, 1965)
Gamma-ray Spectroscopy with Particular Reference to Detector and Computer Evaluation Techniques (P. Quittner, Adam Hilger Ltd, London, 1972)
Instrumentation in Applied Nuclear Chemistry (J. Krugers, Ed., Plenum Press, 1973)
Practical Measuring Techniques for Beta Radiation (L. A. J. Venverloo, Macmillan, 1971)
Introduction to Liquid Scintillation Counting (A. Dyer, Ed., Heyden, 1974)
'Sample preparation procedures for liquid scintillation counting' (B. W. Fox, Chapter 4 in *Specialist Periodical Rpts*, No. 3, 1976)

The statistical approach to counting is important and is discussed in the books listed in the next section.

Basic chemical and handling techniques

A number of textbooks, of which some of the more recent are listed here, cover the basic chemical techniques and also provide information about radioisotopes and counting techniques:

Isotopes in Chemistry (J. F. Duncan and G. B. Cook, Clarendon Press, 1968)

Nuclear and Radiochemistry (G. Friedlander, J. W. Kennedy and J. M. Miller, 2nd edn, Wiley, 1964)

Radio-isotope Laboratory Techniques (R. A. Faires and B. H. Parks, 3rd edn, Newnes, 1973)

Principles of Radiochemistry (H. A. C. McKay, Butterworths, 1971)

Selected References to Tracer Techniques (The Radiochemical Centre, Review 13, 1972)

Isotope User's Guide (F. E. McKinney, S. A. Reynolds and P. S. Baker, ORNL-11C-19, Oak Ridge National Laboratory, Tennessee, 1969)

Users' Guide for Radioactivity Standards (National Academy of Sciences, National Research Council, Washington, DC, 1974)

'Industrial applications of radioisotopes' (J. A. Heslop, Chapter 1 in *Specialist Periodical Rpts*, No. 3, 1976)

'Preparation of radiopharmaceuticals and labelled compounds using short-lived radionuclides' (D. J. Silvester, Chapter 3 in *Specialist Periodical Rpts*, No. 3, 1976)

Radiochemical Survey of the Elements (M. Haissinsky and J. P. Adloff, Elsevier, 1965)

Applications to analytical chemistry

Radioisotopes have found many important applications in the field of analytical chemistry. The types of application are classified as follows:

(1) Method study – a radioisotope employed as a tracer in examining conventional analytical procedures.

(2) Radiometric analysis for the assay of samples which are radioactive or by the application of a radioisotope as a reagent.

(3) Isotope dilution for analysis of complex mixtures where conventional separation procedures are difficult and tedious.

(4) Radio-activation analysis where extreme sensitivity is required; a radioisotope of the test element is produced by neutron activation and the induced activity is a measure of the amount.

Selected examples of applications of the types listed above are as follows:

Introduction to Nuclear Analytical Chemistry (J. Toelgyessy, S. Varga and V. Krivan, University Park Press, Baltimore, 1971)

Treatise on Analytical Chemistry, Part 1, Vol. 9 (I. M. Kolthoff and P. J. Elving, Eds, Wiley, 1971)
Chemical Applications of Radioisotopes (H. J. M. Bowen, Methuen, 1969)
Radiometric Analysis – Methods of Radiochemical Analysis (Fission products in water, foods and animals) (WHO Technical Report Series No. 173; WHO, Geneva, 1966)
Radioactive Isotope Dilution Analysis (The Radiochemical Centre, Amersham)
Isotope Dilution Analysis (J. Toelgyessy, T. Braun and M. Kyrs, Pergamon Press, 1972)
Activation Analysis Handbook (R. C. Koch, Academic Press, 1960)
Guide to Activation Analysis (W. S. Lyon, Van Nostrand, 1964)
Radioactivation Analysis (H. J. M. Bowen and D. Gibbons, Oxford University Press, 1963)
Neutron Activation Analysis (D. DeSoete, R. Gijbels and J. Hoste, Wiley, 1972)
Handbook on Nuclear Activation Cross-Section (IAEA Vienna, 1974)
Activation and Decay Tables of Radioisotopes (E. Bujdoso, I. Feher and G. Kardos, Eds, American Elsevier, 1973)

NUCLEAR CHEMISTRY

The range of radionuclides produced in fission have been the subject of investigations for some years. Reports of the production and characterization of new radioisotopes formed in a variety of ways appear from time to time in *J. Inorg. Nucl. Chem.*, *Phys. Rev.* and *Phys. Rev. Letters*. Topics in nuclear chemistry have received considerable review coverage in the *MTP International Review of Science* and in the Chemical Society *Specialist Periodical Reports* (*SPR*) on Radiochemistry.

'Nuclear fission' (G. N. Walton, Chapter 1 in *MTP Series 1*, 1972)
'Transactinide chemistry' (J. D. Hemingway, Chapter 3 in *SPR* No. 2, 1975)
'Transcurium elements' (R. J. Silva, Chapter 3 in *MTP Series 1*, 1972)
'Superheavy elements' (J. D. Hemingway, Chapter 2 in *SPR* No. 1, 1972; G. Herrmann, Chapter 5 in *MTP Series 2*, 1975)

RADIATION CHEMISTRY

Rather more than 5% of the total energy of nuclear fission appears as delayed radiation from the fission products. Ionizing radiation can initiate or promote a variety of chemical changes, and with the expansion of the nuclear power programme, considerable research into the chemical effects of the radiation and possible applications is in progress. Selected literature is as follows:

Current Topics in Radiation Research (M. Ebert and A. Howard, Eds, North-Holland, 1965)
Latest Soviet Research in Radiation Chemistry (Moscow, 1965)

Atomic Energy Waste; its Nature, Use and Disposal (E. Glueckauf, Ed., Butterworths, 1961)
Introduction to Radiation Chemistry (L. I. Verezhinskii and A. V. Pikaev, Oldbourne Press, 1964)
Radiation Damage in Crystals (L. T. Chadderton, Methuen, 1965)
International Symposium on Pulse Radiolysis (M. Ebert, Ed., Academic Press, 1965)

The chemical effects consequent upon the recoil of a nucleus undergoing a nuclear reaction, e.g. radiative capture (n,γ), have been employed in the preparation of radioisotopes of high specific activity — i.e. the Szilard—Chalmers reaction. Investigations of the purely chemical effects of nuclear recoil have received increasing attention in recent years. Recent review publications have included:

'Chemical effects of nuclear transformations' (G. W. A. Newton, Chapter 1 in *SPR* No. 1, 1972)
'Nuclear recoil chemistry in gases and liquids' (D. S. Urch, Chapter 1 in *SPR* No. 2, 1975; and Chapter 3 in *MTP Series 2*, 1975)
'Recoil chemistry of solids' (P. Glentworth and A. Nath, Chapter 2 in *SPR* No. 2, 1975)
'The chemical effects of nuclear transformations in solids' (A. G. Maddock, Chapter 6 in *MTP Series 2*, 1975)

RADIOLOGICAL HEALTH AND SAFETY

The hazards of ionizing radiation have been known for some time and have been reviewed in:

The Hazards to Man of Nuclear and Allied Radiations: 2nd Report to the Medical Research Council (HMSO, Command 1225, 1960)
The Evaluation of Risks from Radiation (ICRP Publication No. 8, Pergamon Press, 1966)
Harmful Effects of Ionising Radiations (E. Browning, Elsevier, 1959)

Sources of ionizing radiation present a possible twofold hazard:

(1) External radiation from sources outside the body, e.g. sealed sources, X-ray equipment, isotope stores.
(2) Internal radiation from sources deposited inside the body, e.g. radiostrontium in the bone, radioiodine in the thyroid gland.

The Regulations require (*inter alia*) that certain Inspectors appointed by the Health and Safety Executive under Section 19 of the Health and Safety at Work etc. Act 1974 must be notified of the use and disuse of radioactive sources, together with other specific requirements regarding shielding of sources, instruction of employed persons and designation of 'classified workers', whose names must be entered in the Health Register in which the appointed doctor must enter records. Regulations relating to Educational Establishments as set out in The Direct Grant Schools Amending Regulation 1965 (*S.I.* 1965 No. 1), The Further Education Regulations 1969 (*S.I.* 1969 No. 403) and The Schools Amending Regulations 1965 (*S.I.* 1965 No. 3) require that in the interests of health and safety, the approval of the Secretary of State for Education and Science must be obtained for the giving of instruction involving the use of radioactive substances and equipment.

When dealing with sealed sources (e.g. portable neutron sources, X-ray equipment, thickness gauges), the risk is almost entirely confined to external radiation. The industrial handling of such sources is governed by the Factories Act 1961 and The Ionising Radiations (Sealed Sources) Regulations 1969 (*S.I.* No. 808). The industrial application of open sources of radiation where the risks may be internal or external are governed by the Factories Ionising Radiations (Unsealed Radioactive Substances) Regulations 1968 (*S.I.* No. 780).

The maximum permissible levels of internal and external radiation exposure for industrial personnel and members of the public have been laid down in the *Recommendations of the International Commission on Radiological Protection* (ICRP Publication 9, 1966). Maximum permissible doses from external radiation are also set out in the *UKAEA Health and Safety Code No. E.1.1* (HMSO, 1960). Maximum permissible doses from inhaled and ingested radioisotopes are laid down in the *UKAEA Health and Safety Code No. E.1.2* (HMSO, 1961) or alternatively reference may be made to the *Report of the ICRP Committee 2 (1959)*, also published for ICRP by Pergamon Press (1960).

A *Code of Practice for the Protection of Persons Exposed to Ionising Radiations in Research and Teaching* has been prepared by a Panel appointed by the Radioactive Substances Advisory Committee (HMSO, 1968). The *Code* applies to all research laboratories, research establishments and teaching laboratories not subject to the Factories Act, excluding hospital research establishments which have their own code. The Radioactive Substances Act 1960 is designed to ensure effective control over radioactive wastes, and applies to all premises keeping or using radioactive materials. An explanatory memorandum (HMSO, 1963) provides essential information about the Act.

General guidance in the design of facilities and the planning and control of operations with radioactive materials in order to maintain the recommended maximum permissible levels is available in a number of publications:

IAEA Publications in Health and Safety — Safety Series

Protection against X-rays up to Energies of 3 MeV and Beta-Gamma Rays from Sealed Sources (ICRP Report of Committee III, Pergamon Press, 1960)
Principles of Radiation Protection, Health Physics (D. J. Rees, Butterworths, 1967)
Principles of Radiation Protection (K. Z. Morgan and J. E. Turner, Eds, Wiley, 1967)
Basic Radiation Protection Criteria (NCRP-39. National Council on Radiation Protection and Measurements, Washington, DC, 1971)
Nuclear Safeguards (McGraw-Hill, 1974)

A summary of the law relating to atomic energy and radioactive substances in the UK is available from UKAEA. Many countries have their own domestic laws which are not applicable elsewhere. A survey of the current world legislation as at November 1971 is available (*Protection against Ionizing Radiation*, WHO, Geneva), while a comparative study of the major aspects of nuclear energy legislation in force in OECD Member countries has been published by the Nuclear Energy Agency (obtainable from HMSO).

APPENDIX: USEFUL ADDRESSES

EURATOM	EURATOM Transatom Service, 51 rue Belliard, Brussels, Belgium
IAEA	IAEA INIS Section, International Atomic Energy Agency, P.O. Box 590, A-1011 Wien, Austria
The Radiochemical Centre	The Radiochemical Centre, Amersham, Buckinghamshire, HP7 9LL, England
UKAEA	UKAEA Public Information Centre, 11 Charles II Street, London SW1Y 4QP, England
WHO	World Health Organization, 1211 Geneva 27, Switzerland

11

Beilstein's 'Handbuch' as a source of information on organic chemistry

T. C. Owen and *R. M. W. Rickett*
(revised by *R. T. Bottle*)

NATURE, USES AND LIMITATIONS

The *Handbuch der organischen Chemie*, initiated by Friedrich Konrad Beilstein,* is the largest compilation of information on organic chemistry. It gives an exhaustive survey of organic compounds which have been definitely characterized and also of naturally occurring organic materials which are of unknown structure. It is, therefore, a most voluminous work; in the current (fourth) edition, the main work (*Hauptwerk, HW*) comprises some 200 000 entries in 27 volumes covering the literature prior to 1910, and this is augmented by supplementary series of volumes, each corresponding in content to the conumerate volume in the *Hauptwerk*. The first supplement (*Erstes Ergänzungswerk, EI*) covers the literature from 1910 to 1919 and the second (*Zweites Ergänzungswerk, EII*) from 1920 to 1929.

In 1958 the first volumes of the third supplement (*Drittes Ergänzungswerk, EIII*), covering the literature from 1930 to 1949, appeared,

---

*Beilstein, a Russian who had studied in Germany, completed the first edition of his *Handbuch* in 1882, and subsequently produced two further editions. The fourth edition (and also supplements to the third edition, 1899–1906) were issued by the German Chemical Society until after World War II, when its preparation was vested in the Beilstein Institut, Frankfurt a/M. For further historical details, see E. H. Huntress, *J. Chem. Educ.*, **15**, 303 (1938).

and in 1972 volumes of the fourth supplement (*Fiertes Ergänzungswerk, E IV*), covering the literature from 1950 to 1959, started to appear. As from Vol. 17, these were combined into a single supplement (*E III/IV*) with literature coverage from 1930 to 1959. In the more recently published volumes literature coverage frequently goes beyond the cut-off date and extends to within about four years of the publication date.

Beilstein is now an extremely expensive serial, each 'volume' consisting of several parts averaging around £300 each part. Second-hand sets are often available from librarians unwilling to continue this commitment, and microform versions of the main work and first two supplements are available more cheaply.

The unique and important features of *Beilstein* are its *comprehensiveness* (information on all known compounds being listed) and its *compilation* of the information. In contrast, the comprehensive information in *Chemical Abstracts* is scattered, while the compiled information in the various dictionaries is far from complete.

Thus, a comprehensive survey of the literature (prior to 1950) on ethyl benzoate simply involves a few pages of Vol. 9 of *HW, EI, EII* and *EIII*; extraction of the same information from other sources is a major operation involving considerable mental (and physical!) labour. It is our experience that information is often missed in such a search. However, *Beilstein* can never be up to date. No supplementary volume can ever appear until the whole of the period which it covers has elapsed and in practice a long delay has always occurred. *EII* was not completed until almost thirty years after the end of 1929. The work, therefore, gives no guide to the recent literature. It is an information retrieval system for ascertaining the major physical and chemical properties, preparations, etc., of any organic compound reported in the literature during the period covered. It also gives a detailed guide to the extent of knowledge in a given field together with ample references to the original literature. The need to consult the early abstract literature, which may be difficultly accessible, incomplete and sometimes even misleading, is thus removed. The information so obtained is frequently sufficient for one's requirements, more recent data often being unnecessary.

INFORMATION LISTED

The information listed varies from a few lines to many pages, depending on the importance of the particular compound. A full but abbreviated account is given, including, as available, name(s); formula;

structure(s); history; occurrence; formation; preparation; physical, chemical and physiological properties; technology; analysis; addition compounds; salts; and conversion products of unknown structure.

FINDING A COMPOUND IN BEILSTEIN

Indexes

The location of information in *Beilstein* has been considerably simplified since the cumulated general indexes to the *Hauptwerk*, first and second supplements, were published in 1956.

Location by name is alphabetical in the subject index (*Sachregister*; *EII*, Vol. 28, parts 1 and 2). Location in the formula index (*Formelregister*; *EII*, Vol. 29, parts 1, 2 and 3) is according to the Hill system (see page 70). The entry against each formula is subdivided according to the names of the isomeric substances, each of which is followed by the appropriate volume and page numbers. Thus the entry **20**, 181, I 54, II 96, **22** II 619 against *pyridine* in the subject index and against C_5H_5N–pyridine in the formula index indicates data on this compound on page 181 of Vol. 20 in the *Hauptwerk*, page 54 of the same volume in the first supplement, page 96 of this volume in the second supplement, and page 619 of Vol. 22 in the second supplement. Similar general indexes to *HW* and *EI* only were published in 1938–40 as Vols 28 and 29 of the *Hauptwerk*. The only value these now have is in a few cases where the complete work including the 1956 indexes may not be available. Care should be exercised in case the earlier indexes should be consulted in mistake for the later set. Furthermore, it should be noted that while the 1956 indexes use the Hill classification system, earlier formula indexes use the Richter system, in which other elements follow C and H in a set sequence (O, N, Cl, Br, I, F, S, P). In addition to the general indexes, each volume has its own subject and formula index. These are of value in cases where the general indexes are anteceded (e.g. *EIII and EIV*).

The various guides to the use of *Beilstein* (see Bibliography) throw great emphasis on an understanding of the system of classification adopted in the construction of the work as a means of locating information. This is because general indexes did not exist prior to 1938, and only to the *Hauptwerk* and first supplement prior to 1956. Between these dates it was possible to locate entries in the second supplement for compounds previously reported in the earlier series by either of the following methods: the appropriate earlier volume number could be determined and the index to the appropriate volume in the second supplement consulted or the system number (*Systemnummer*) for the compound could be observed in the earlier work and the same number located in the volume of the later series. The system numbers comprise a series of 4877 arbitrary units into which the subject matter is subdivided and one can quickly find a specific compound by scanning the entries under the appropriate system number. Data on compounds first reported after 1920 could not be located in this way, and a knowledge of the classification system adopted by the editors was essential.

Compounds noted in earlier series may be located in the third and fourth supplements by utilizing the corresponding main work pagination (e.g. H 123), which is centrally printed at the top of each page. Other compounds can often be similarly located by looking for the pagination of closely related compounds. If this method fails, one must then locate the compound by means of the system as described below.

Table 11.1 FUNCTIONING CLASSES IN THE BEILSTEIN CLASSIFICATION SYSTEM

---

1. Stem nuclei (and non-functioning derivatives)
2. Hydroxy compounds
3. Carbonyl compounds
4. Carboxy acids
5. Sulphinic acids
6. Sulphonic acids
7. Seleninic and selenonic acids
8. Amines
9. Hydroxylamines
10. Hydrazines
11. Azo compounds
12. Hydroxyhydrazines
13. Diazonium, diazo and isodiazo compounds
14. Azoxy compounds
15. Nitramines and isonitramines
16. Triazanes
17. Triazenes
18. Hydroxytriazenes
19. Azoamino-oxides
20. Tetrazanes
21. Tetrazenes
22. Compounds with chains of more than four nitrogen atoms
23–28. Organometallic compounds having carbon bonded directly to the metallic element, *except* salts and all compounds of the alkali metals.

(If one memorizes the order of the groups in Classes 1, 2, 3, 4, 6 and 8, in particular, this will prove most useful.

Table 11.2 CORRESPONDENCE OF VOLUMES AND SYSTEM NUMBERS
WITH DIVISIONS AND CLASSES

Division	Sub-division	Class	Sub-class	Volume	System number
I	–	1–3		I	1–151
I	–	4a		II	152–194
I	–	4b	Polyfunctional carboxy acids	III	195–322
I	–	5 and over		IV	323–449
II	–	1		V	450–498
II	–	2		VI	499–608
II	–	3a	Mono- and poly car-bonyl compounds	VII	609–736
II	–	3b	Hydroxycarbonyl compounds	VIII	737–890
II	–	4a	Mono- and polycarboxy acids	IX	891–1050
II	–	4b	Hydroxy and carbonyl acids	X	1051–1504
II	–	5–7		XI	1505–1591
II	–	8a	Monoamines	XII	1592–1739
II	–	8a and b	Poly- and hydroxy amines	XIII	1740–1871
II	–	8b	Other amines	XIV	1872–1928
II	–	9 and 10		XV	1929–2084
II	–	11 and over		XVI	2085–2358
III	1 cyclic O (S, Se, Te)	1–3a		XVII	2359–2503
III	1 cyclic O (S, Se, Te)	3b and over		XVIII	2504–2665
III	2 or more cyclic O	All		XIX	2666–3031
III	1 cyclic N	1		XX	3032–3102
III	1 cyclic N	2 and 3		XXI	3103–3241
III	1 cyclic N	4 and over		XXII	3242–3457
III	2 cyclic N	1 and 2		XXIII	3458–3554
III	2 cyclic N	3a		XXIV	3555–3633
III	2 cyclic N	3b and over		XXV	3634–3793
III	More than 2 cyclic N	All		XVI	3794–4187
III	Cyclic O and N	All		XXVII	4188–4720
	Subject index. Part I, A–G; Part II, H–Z			XXVIII	
	Formula index. Three parts			XXIX	

In 1922 the following volumes appeared for which there are no supplements.

IV Rubber, Gutta Percha, Carotenoids				XXX	4723–4723a
IV Carbohydrates				XXXI	4746–4767a

NOTE: 'a' indicates unmixed functionality; 'b' indicates mixed functionality.

Use of the classification system

Any substance of known structure is entered at a position which is absolutely determined by that structure. Each structural formula is therefore its own index. Location is effected by applying the following three operations.

(1) Examination of the structural formula (with manipulations as detailed below) determines to which division, sub-division, class, sub-class, etc., the substance belongs (cf. *Table 11.1*).

(2) Inspection of *Table 11.2* and application of the classification thus determined indicates the appropriate volume.

(3) Random opening of the appropriate volume and examination of the entries on the page so selected will make it immediately apparent whether the desired compound must appear before or after the entries inspected (provided that one knows the classification scheme). Adjustment and repetition of the inspection process soon leads to location of the desired compound.

Perusal of a worked example will clarify the application of these operations. The example should be re-examined in detail when the later section on the classification system has been studied.

Example. 4-Chloro-6-hydroxypyridine-2-carboxylic acid.

Inspection shows this to belong to Division III (heterocyclic), subdivision one cyclic nitrogen, latest 'functioning group' carboxylic acid, Class 4. *Table 11.2* therefore indicates Vol. 22. Random opening of this volume, at page 443, for example, reveals entries concerning aminoquinolines. Clearly this is the correct division and sub-division, but amines (Class 8) come later than the desired carboxylic acids, so pages are skimmed through in the direction of the front of the book until heterocyclic carboxylic acids appear (page 385). Again stopping at random, page 133 reveals dicarboxylic acids. These must come before hydroxycarboxylic acids, so pages are checked, somewhat more carefully, in the reverse direction until hydroxycarboxylic acids with three atoms of oxygen appear at page 190. Close inspection then reveals successively hydroxypyridine carboxylic acids, page 212, under which is 6-hydroxypyridine-2-carboxylic acid, page 213, and the desired 4-chloro-6-hydroxypyridine-2-carboxylic acid as a 'stem nuclear non-functioning derivative' thereof on page 214.

THE CLASSIFICATION SYSTEM

Divisions

The four divisions comprise:

Division I–*Acyclic*
Division II–*Alicyclic*
Division III–*Heterocyclic*
Division IV–*Natural products not assigned in the above three divisions*

The assignment of a compound to Division I, II or III depends on the derivation of its 'stem nucleus', which is obtained by replacing all atoms or groups attached to carbon by the appropriate number of hydrogen atoms, except where such replacement would require rupture of a heterocyclic ring.

Example

Compound	Stem nucleus	Division
$CH_2ClCH_2CH_2COOH$	$CH_3CH_2CH_2CH_3$	I
$CH_3CH_2COOCH_2CH_3$	$CH_3CH_2CH_3$ and CH_3CH_3	I
$C_4H_9SO_2SC_2H_5$	C_4H_{10} and C_2H_6	I

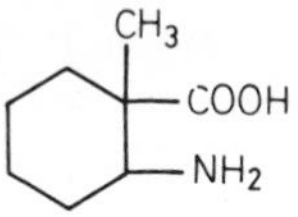

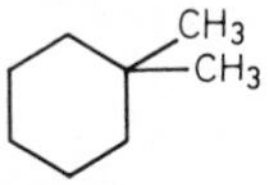

II

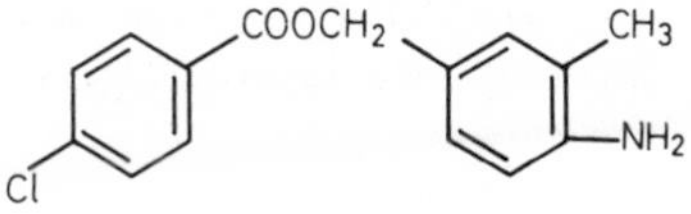

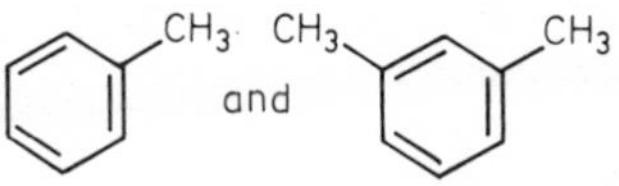

II

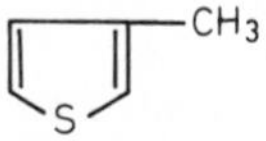

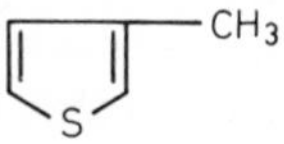

III

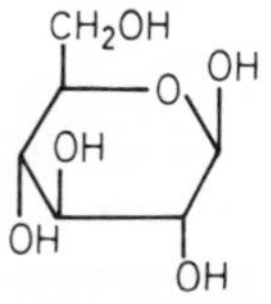

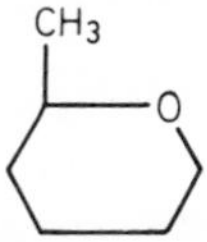

III

Example

Compound	Stem nuclei	Division
$CH_3CH_2COO-\langle\text{ring}\rangle$	$CH_3CH_2CH_3$ and $\langle\text{ring}\rangle$	II
NO, COOCH₂, pyridine with NO₂, Br substituents	$\langle\text{ring}\rangle-CH_3$ and $CH_3-\langle\text{pyridine}\rangle$	III

Frequently, stem nuclei corresponding to more than one division occur in one compound. In such cases the compound is assigned to the division corresponding to that stem nucleus which comes last in the classification. This *principle of latest position* is applied widely elsewhere in the classification system.

Sub-divisions

The stem nuclei of Divisions I and II contain only carbon and hydrogen, so that further classification depends on the type and number of functional groups present (see 'Functioning classes', below). In Division III, however, other atoms are present in the stem nucleus, necessitating the definition of a sequence of *sub-divisions* governing the order in which heterocyclic stem nuclei must be placed before further classification according to functional groups may be considered. The defined sequence considers first oxygen and then nitrogen heterocycles as follows:

(a) stem nuclei containing $1,2,3,4,...n$ oxygen atoms;
(b) stem nuclei containing $1,2,3,4,...n$ nitrogen atoms;
(c) stem nuclei containing $1,2,3,...n$ oxygen and 1 nitrogen;
(d) stem nuclei containing $1,2,3,...n$ oxygen and 2 nitrogen, etc.

Heterocyclic stem nuclei containing the heavier elements of group 6 of the periodic table, sulphur, selenium and tellurium, are considered to arise by replacement of oxygen and are listed immediately following the parent stem nuclei containing that element. Heterocycles containing other cyclically bound atoms are considered after all O, N, S, Se,

Te heterocyclics have been dealt with. In cases where stem nuclei corresponding to more than one sub-division occur in one compound, assignment is to the last possible sub-division in accordance with the principle of latest position in the system.

Functioning classes

Having determined the gross order by division and sub-division according to stem nuclei, the order of consideration is next defined according to the functional groups attached thereto. The order of precedence is detailed in *Table 11.1*, the order of the first eight or ten classes therein being of overriding importance. Stem nuclear compounds having either no substituents or the seven 'non-functioning' substituents* $-F, -Cl, -Br, -I, -NO, -NO_2, -N_3$ only comprise Class 1 and are dealt with in order of decreasing saturation and increasing number of carbon atoms. Thus Vol. 1 deals with Division I, Class 1 in the order:

Hydrocarbons C_nH_{2n+2}

Methane and its non-functioning derivatives;
ethane and its non-functioning derivatives;
propane and its non-functioning derivatives;
n-butane and its non-functioning derivatives.
Similarly isobutane, n-pentane, dimethylethylmethane, tetramethylmethane, n-hexane, ... higher paraffins, each with its non-functioning derivatives.

Hydrocarbons C_nH_{2n}

Ethylene and its non-functioning derivatives;
propylene and its non-functioning derivatives;
1-butene and its non-functioning derivatives.
Similarly 2-butene, isobutylene, 1-pentene, etc.

In an analogous manner, Vol. 5 deals with Division II, Class 1 in the order:

Hydrocarbons C_nH_{2n}

Cyclopropane and its non-functioning derivatives.
Cyclobutane, methyl cyclopropane, cyclopentane, methyl cyclobutane, 1,1-dimethyl cyclopropane, etc., each with its non-functioning derivatives.

*These contain no replaceable hydrogen atoms.

Hydrocarbons C_nH_{2n-2}

> Cyclopropene, cyclobutene, cyclopentene, methyl cyclobutene, methylene cyclobutane, etc., each with its non-functioning derivatives, etc.

After all known compounds having no functioning groups* have been entered in each division, those containing such groups are dealt with in order as in *Table 11.1*. The functioning groups of classes 2 to 22 all contain replaceable hydrogen and are joined to the stem nucleus by single bonds, carbonyl compounds being

regarded in this connection as *gem*-diols,

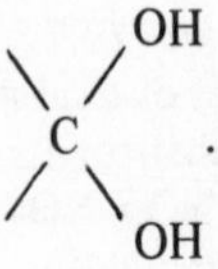

$$\begin{array}{c} \diagdown \quad \diagup OH \\ C \\ \diagup \quad \diagdown OH \end{array}$$

. Compounds con-

taining a functioning group attached to a carbon atom which bears one or more other functioning or non-functioning groups of any kind are regarded as 'replacement derivatives' (see below). They appear under that carbonyl compound or carboxylic acid which results from replacement of all these groups by an appropriate number of hydroxyl groups followed by elimination of water from *gem*-diols as may be appropriate. The same applies to compounds in which nitrogen is joined to carbon by multiple valence.

Examples

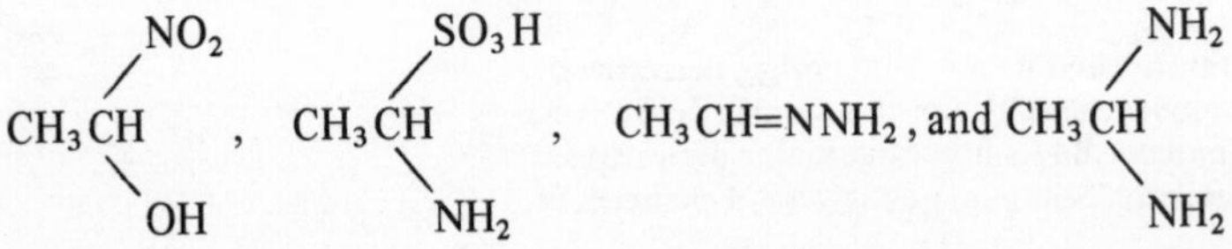

$$\begin{array}{c} \diagup NO_2 \\ CH_3CH \\ \diagdown OH \end{array}, \quad \begin{array}{c} \diagup SO_3H \\ CH_3CH \\ \diagdown NH_2 \end{array}, \quad CH_3CH{=}NNH_2, \text{ and } \begin{array}{c} \diagup NH_2 \\ CH_3CH \\ \diagdown NH_2 \end{array}$$

would be all classed as derivatives of, and would be considered under

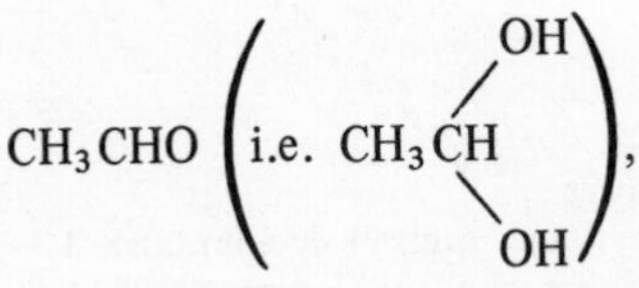

$$CH_3CHO \left(\text{i.e. } \begin{array}{c} \diagup OH \\ CH_3CH \\ \diagdown OH \end{array} \right), \text{ Division I, Class 3.}$$

*Functioning must not be confused with the common term *functional*.

$$CH_3CH \begin{matrix} \diagup NO \\ \diagdown Cl \end{matrix} \quad \text{is a derivative of } CH_3CH_3, \text{ Division I, Class 1.}$$

$CH_3CH_2NHNH_2$ and $CH_3CH_2SO_3H$ are functioning compounds of Division I, Classes 10 and 6, respectively.

It frequently occurs that different functioning or non-functioning groups occur, or may be conceived to occur, on different carbon atoms of a given stem nucleus. The principle of latest position is applied in such cases. Any non-functioning group, therefore, is considered after all derivatives of the functioning groups present, e.g. 4-chloroanthranilic comes after all the esters, amides, etc., of anthranilic acid.

Examples

Bromoacetic acid is considered under acetic acid, Division I, Class 4, Vol. 3, as a non-functioning derivative, just as ethyl bromide is considered under ethane.

Ethanolamine, $HOCH_2CH_2NH_2$, is considered under ethylamine, Division I, Class 8, Vol. 4.

Where several functioning groups occur, the latest functioning group is considered first, *followed by the others in the order in which they appear in Table 11.1*. Thus the order of appearance of the classes in any division (or sub-division of Division III), non-functioning derivatives being considered along with each parent functioning compound, would be:

Stem nuclear compounds
Monohydroxy compounds
Dihydroxy compounds
Trihydroxy compounds,
    etc.
Monocarbonyl compounds
Dicarbonyl compounds,
    etc.
Carbonyl–hydroxy compounds –with 2 atoms of oxygen
                                     – with 3 atoms of oxygen
                                     – with 4 atoms of oxygen,
                                       etc.
Monocarboxy compounds
Dicarboxy compounds
Tricarboxy compounds,
    etc.

Carboxy–hydroxy compounds – with 3 atoms of oxygen
– with 4 atoms of oxygen
– with 5 atoms of oxygen,
etc.
Carboxy–carbonyl compounds – with 3 atoms of oxygen
– with 4 atoms of oxygen,
etc.
Carboxy–hydroxy–carbonyl compounds – with 4 atoms of oxygen
– with 5 atoms of oxygen,
etc.
Monosulphinic acids
Disulphinic acids, etc.
Hydroxy–sulphinic acids – with 1 OH and 1,2,3,4, etc., SO_2H groups
– with 2 OH and 1,2,3,4, etc., SO_2H groups, etc.
Carbonyl–sulphinic acids – with 1 C=O and 1,2, etc., SO_2H groups
– with 2 C=O and 1,2, etc., SO_2H groups, etc.
Carbonyl–hydroxy–sulphinic acids – with 2 atoms of oxygen in the C=O
and 1,2, etc., SO_2H groups
– with 3 atoms of oxygen and 1,2,3,
etc., SO_2H groups, etc.
Carboxy–sulphinic acids – as for hydroxysulphinic acids
Carboxy–hydroxy–sulphinic acids – with 3 atoms of oxygen and 1,2,3,
etc., SO_2H groups
– with 4 atoms of oxygen, similarly,
etc.
Carboxy–carbonyl–sulphinic acids – as for carboxy–hydroxy
Carboxy–hydroxy–carbonyl–sulphinic acids – similarly

While the number of variations obtainable in this manner is as limitless as the number of organic compounds which it is possible to envisage, it is obvious that the majority of compounds corresponding to the higher combinations will not be known. However, from a knowledge of the order in which the combinations will appear, it is possible to judge whether the entries on a page, selected by opening the appropriate volume at random, precede or antecede the desired compound. Repetition of the process leads to easy location of the desired entry. With a little practice, location by this means becomes easily as rapid and convenient as the use of the indexes, and is, of course, obligatory where indexes are not available.

Further subclassification

Application of the classification system as above is invariably sufficient to locate any known compound quite easily. For completeness, it may be mentioned that each of these groups (e.g. acyclic monohydroxy compounds) is arranged in order of decreasing saturation (C_nH_{2n+2}, C_nH_{2n}, C_nH_{2n-2}, etc.), and then each group having a given degree of saturation (acyclic monohydroxy compounds, C_nH_{2n}) is grouped in order of increasing number of carbon atoms.

Functioning derivatives

Functioning derivatives are those compounds which arise formally by 'coupling' a functioning group with a 'coupling component'. In such coupling there is eliminated as water an H from the functioning group and an OH of an organic coupling component, or an H or OH from the functioning group and an OH or H of an inorganic coupling component. Such derivatives are located by 'uncoupling' followed by location of the parent functioning compounds.

A discussion of the order of precedence of such derivatives is beyond the scope of the present work and is unnecessary for the location of a given derivative. However, when both the parent functioning compound and the coupling component are organic, ambiguity frequently arises as to which is which. The principle of latest position is applied, the compound being regarded as a derivative of that one of its components which comes latest in the system.

Examples

(1) *Ethyl benzoate.* Uncoupling gives ethanol and benzoic acid, which can be recoupled in two ways regarding either as the parent functioning compound and the other as the coupling component. Benzoic acid, however, comes later in the system than ethanol, so that this ester is regarded as a derivative of the acid.

(2) *Benzyl acetate.* By a similar process, and in contrast to the previous example, this ester is regarded as a derivative of benzyl alcohol, with acetic acid as the coupling component, the alicyclic (Division II) alcohol coming later than the acyclic acid.

(3) N-*Ethyl*-p-*chlorobenzamide*, $ClC_6H_5CONHCH_2CH_3$. Whereas in examples (1) and (2) uncoupling on either side of the singly bonded ester oxygen gave the same components, in this case uncoupling the CO—N bond gives p-chlorobenzoic acid and ethylamine, while uncoupling the N—CH bond gives p-chlorobenzamide and ethanol. p-Chlorobenzamide is not an index compound and must be further uncoupled, giving ammonia and again p-chlorobenzoic acid. The latest of these is the acid, and it is under this compound that the substituted amide will be listed — not, however, as a derivative with ethylamine, which would require elimination of H instead of OH from an organic coupling component, but as a second-degree derivative from coupling with ammonia, elimination of H from inorganic coupling components being permitted, and then with ethanol.

Non-functioning and replacement derivatives of functioning compounds and functioning derivatives

These are obtained by substitution of stem nuclear hydrogen by non-functioning substituents and of functioning oxygen by sulphur, selenium or tellurium. They and their derivatives appear after the parent functioning compounds and their functioning derivatives have all been disposed of. Thus ethyl chloracetate is considered as a functioning derivative of chloracetic acid, which is itself a non-functioning derivative of acetic acid and comes after that acid and all its functioning derivatives have been dealt with. The ester is not regarded as a derivative of ethyl acetate.

With some practice, location of compounds by use of this system becomes easy. The above is, however, an abbreviated guide and in a few instances may not suffice. In such cases it is most instructive to locate the desired compound, or something closely related to it, by means of indexes and then to examine the surrounding text, whereupon the factors which decide its location are determinable. If further help is needed in locating compounds in *Beilstein*, use Runquist's programmed guide or Weissbach's recent guide.

BIBLIOGRAPHY

Beilstein's *Handbuch der organischen Chemie* (1918). 4th edn, Vol. I, Springer, pp. 1–47
System der Organischen Verbindungen: Ein Leitfaden für die Benutzung von Beilsteins Handbuch (1929). Deutsche Chemische Gesellschaft, Springer
Huntress, E. H. (1938). *A Brief Introduction to the Use of Beilstein's Handbuch*, 2nd edn, Wiley
Runquist, O. (1966). *Beilstein's Handbuch, a Programmed Guide*, Burgess
Hancock, J. E. H. (1968). *J. Chem. Educ.*, 45, 336
Weissbach, O. (1976). *The Beilstein Guide* (translated by H. M. R. Hoffmann), Springer

12

Other reference works on organic chemistry

A. G. Osborne

There has been a considerable increase in the volume of published work on organic chemistry since the last edition of this book was written. Consequently, the range of reference works, serials and data compilations has also increased, particularly in the area of regular reviews and updates of the literature.

Organic Chemistry (A. T. Blomquist, Ed., 1964–71; A. T. Blomquist and H. Wasserman, Eds, Academic Press, 1971–   ) is a series of monographs on topics in organic chemistry; certain volumes have appeared with later editions. The following volumes have appeared so far:

1, Carbene chemistry; 2, Bridged aromatic compounds; 3, Conformation theory; 4, Carbanion chemistry; 5, Oxidation in organic chemistry; 6, Structure and mechanism in organophosphorus chemistry; 7, Ylid chemistry; 8, 1,4-Cycloaddition reactions; 9, Cycloaddition reactions of heterocumulenes; 10, Cyclobutadiene and related compounds; 11, Dehydrobenzene and cycloalkynes; 12, Organic functional group preparations; 13, Ring-forming polymerisations, parts A and B; 14, Carboxylic ortho acid derivatives; 15, Organic charge-transfer complexes; 16, Nonbenzoid aromatics, Vols I and II; 17, Acidity functions; 18, Chemistry of indoles; 19, Chemistry of the heterocyclic *N*-oxides; 20, Isonitrile chemistry; 21, Conformational analysis; 22, Catenanes, rotaxanes and knots; 23, Reaction mechanisms in sulphuric acid and other strong acid solutions; 24, Carbon-13 NMR spectroscopy; 25, The isoquinoline alkaloids: chemistry and pharmacology; 26, Organic reactive intermediates; 27, Ring transformations of heterocycles, Vols I and II; 28, Organic synthesis with noble metal catalysts; 29, Polymer synthesis, Vol. I; 30, Total synthesis of steroids; 31, Sulphur ylides: emerging synthetic intermediates; 32, Anodic oxidation; 33, Transition metal organometallics in organic synthesis.

Chemistry of Carbon Compounds (E. H. Rodd, Ed., 1st edn, Elsevier, 1951–62; *Rodd's Chemistry of Carbon Compounds* (S. Coffey, Ed., 2nd edn, Elsevier, 1964–   ); *Supplements to the 2nd edition (Editor S. Coffey) of Rodd's Chemistry of Carbon Compounds* (M. F. Ansell, Ed., Elsevier, 1973–   ). '*Rodd*' is a general introduction to organic chemistry and is complete in five volumes, each volume being in several parts. A General Index is published as Vol. 5. A valuable theoretical interpretation of the properties and reactivity of each group of compounds is given in addition to the structural formula, methods of preparation, and chemical and physical properties of the individual compounds listed. Publication of a completely rewritten second edition began in 1964; the first part of Vol. 4, which is to appear in at least ten parts, was published in 1973. Owing to the very long time involved in producing the volumes in this series, a range of supplements (edited by M. F. Ansell) were commenced in 1973, and to date Vols 1 and 2 have been covered.

Elsevier's Encyclopaedia of Organic Chemistry (F. Radt, Ed., Elsevier, 1940–56; F. Radt, Ed., Springer, 1959–69). Only Vols 12, 13 and 14, dealing with condensed carboisocyclic compounds, had appeared when publication was suspended with the fourth supplement to Vol. 14 in 1956. However, four further supplements have been published by Springer to bridge the gap until the steroid sections of *Beilstein* appear. The complete work deals with bicyclic, tricyclic, tetracyclic and higher cyclic compounds, including triterpenes and steroids, and contains separate surveys of methods for the synthesis of the more important ring systems in addition to information on individual compounds. It will be a useful complement to *Beilstein* until the volumes containing the more recent information on polycyclic compounds appear.

Dictionary of Organic Compounds (I. Heilbron, H. M. Bunbury *et al.*, Eds, 4th edn, Eyre and Spottiswoode, 1965). The new, revised and enlarged 4th edition in five volumes was published with the first supplement in 1965 and covers the literature up to the end of 1964. Supplements to revise and extend the main work are published, with cumulative supplements at five-yearly intervals. The fifth and cumulative supplement, covering the literature up to the end of 1968, was published in 1969, and the tenth and cumulative supplement, covering the literature up to the end of 1973, was published in 1974. The fourteenth supplement (literature of 1977) appeared in 1978. A Formula Index covering the main 4th edition and the first five supplements appeared in 1971; the sixth and subsequent supplements include their own formula index.

The formula, source, some physical constants and chemical properties and literature references to preparation are given for each parent compound, together with a list of derivatives. Nomenclature in the new edition conforms to IUPAC recommendations, but continuity with the earlier editions has been preserved by use of cross-references, and many trade names (e.g. of drugs and insecticides) have been included.

Faraday's Encyclopaedia of Hydrocarbon Compounds (J.E. Faraday *et al.*, Eds, Chemindex/Butterworths, 1945–69). Thirteen volumes of the encyclopaedia have appeared, in loose-leaf ring-binder format. Each volume contains the original compilation of data and numerous supplements. Detailed information on the synthesis, physical properties and analysis of hydrocarbons containing 1–14 carbon atoms is given.

Handbook of Organometallic Compounds (N. Hagihara, M. Kumada and R. Okawara, Eds, Benjamin, 1968). This handbook gives details of the physical and chemical properties, synthesis and reactions of organometallic compounds. It is arranged by metallic element or group of elements and each compilation of data is preceded by a general introduction on the group of metals.

Chemistry of Functional Groups (S. Patai, Ed., Wiley, 1964–). This series is intended to cover all aspects of the chemistry of the important functional groups in organic chemistry. Coverage is generally restricted to the more recent developments and to subjects inadequately covered by reviews. Several more than twenty volumes originally planned have been published so far. Representative titles include: *Chemistry of Alkenes* (2 vols, 1964); *Chemistry of the Ether Linkage* (1967); *Chemistry of the Hydrazo, Azo and Azoxy Groups* (2 parts, 1975).

Reactive Intermediates in Organic Chemistry (G. A. Olah, Ed., Wiley, 1968–). A series of volumes, un-numbered, dealing with reactive intermediates in organic chemistry. The following have so far appeared: *Carbonium Ions* (4 vols, G. A. Olah and P. von R. Schleyer, 1968–73, Vol. 5 in press); *Radical Ions* (E. T. Kaiser and L. Kevan, 1968); *Nitrenes* (W. Lwowski, 1970); *Carbenes* (2 vols, R. A. Moss and M. Jones Jr., 1973–75); *Free Radicals* (2 vols, J. K. Kochi, 1973); *Halonium Ions* (G. A. Olah, 1975). Volumes dealing with Carbanions and with Arynes are planned for the series.

Reaction Mechanisms in Organic Chemistry (Init. Ed. C. D. Hughes, later C. Eaborn and N. B. Chapman, Eds, Elsevier, 1963–). The following monographs in this series have appeared: 1, *Nucleophilic*

Substitution at a Saturated Carbon Atom (1963); 2, *Elimination Reactions* (1963); 3, *Electrophilic Substitution in Benzenoid Compounds* (1965); 4, *Electrophilic Additions to Unsaturated Systems* (1966); 5, *The Organic Chemistry of Phosphorus* (1967); 6, *Aromatic Rearrangements* (1967); 7, *Steroid Reaction Mechanisms* (1968); 8, *Aromatic Nucleophilic Substitution* (1968); 9, *Carbanions: Mechanistic and Isotopic Aspects* (1975); 10, *Mechanisms of Oxidation by Metal Ions* (1976).

Conformational Analysis (E. L. Eliel, N. L. Allinger, S. J. Angyal and G. A. Morrison, Wiley, 1965). A treatment of conformational analysis which includes both basic principles and applications to natural product chemistry. Also useful is the *Topics in Stereochemistry* series edited by Allinger and Eliel, which first appeared in 1967. This series contains authoritative monographs in which recent advances in stereochemistry are discussed.

The Essential Oils (E. Guenther, Ed., Van Nostrand, 1948–52). This work is complete in six volumes, the first two containing data on the chemical groups which comprise the major constituents of the essential oils. Each botanical family is given one chapter and the botany, geographical distribution and cultivation of the plant sources are described in addition to the distillation, physical and chemical properties, production and industrial uses of the essential oils. The work contains a large number of references to other sources of information.

Friedel–Crafts and Related Reactions (4 vols, G. A. Olah, Ed., Interscience, 1963–65); *Friedel–Crafts Chemistry* (G. A. Olah, Wiley, 1973). These summaries give experimental conditions as well as general theory and reaction mechanisms for the alkylation, acylation, isomerization and polymerization reactions which fall under this heading. The later supplementary volume contains reprints of the general review chapters of the original volumes with some later additions as well as new chapters on recent general advances.

Organic Sulphur Compounds (N. Kharasch, Ed., Vol. 1, Pergamon Press, 1961); *Chemistry of Organic Sulphur Compounds* (N. Kharasch and C. Y. Meyers, Eds, Vol. 2, Pergamon Press, 1966). A series of well-documented articles on organic sulphur compounds which includes chapters on the biochemical functions of selected compounds and studies on reaction mechanisms in addition to the synthesis and reactions of the more important groups of these compounds. Further volumes in the series were planned but none have yet appeared.

The Chemistry of Heterocyclic Compounds (A. Weissberger, Ed., 1950–70, A. Weissberger and E. C. Taylor, Eds, Interscience, 1970–). This set of monographs is intended to cover comprehensively the chemistry of the complete field of heterocyclic chemistry. It has been appearing at regular intervals since 1950; Vol. 29 on benzofurans, by A. Mustafa, appeared in 1974. The series also includes several multi-volume works and the four-part work on pyridine and its derivatives (Vol. 14), edited by E. Klingsberg (1960–64) and the supplement, edited by R. A. Abramovitch (1974–75), could be considered as a series in its own right. Several volumes have revised editions or supplements.

Heterocyclic Compounds (R. C. Elderfield, Ed., Wiley, Vol. 1, 1950, Vol. 9, 1967). This series is obviously less comprehensive than the much larger Weissberger but is nevertheless very useful. Each chapter discusses the chemistry of a group of heterocyclic compounds under the main headings of synthesis and reactions. The chapters are then grouped into volumes by general heterocyclic types – e.g. Vol. 1: 3-, 4-, 5- and 6-membered polycyclic compounds containing one O, N or S atom; Vol. 2: 5- and 6-membered polycyclic compounds containing one O or S atom.

The Chemistry of Natural Products (K. W. Bentley, Ed., Interscience, 1957–65). A series of monographs on topics in organic chemistry intended mainly for undergraduates, it is noted for the liberal use of structural formulae and reaction schemes, which are separated from the text; original literature references are given. Fields covered include alkaloids, carbohydrates, terpenes, vitamins and natural pigments.

Natural Products Chemistry (2 vols, K. Nakanishi, T. Goto, S. Ito, S. Natori and S. Nozoe, Eds, Academic Press, 1974–75). This work attempts to bridge the gap between the organic textbook and the comprehensive treatise. Structure, syntheses and reactions of classes of natural products and of individual members are given. The presentation uses many structural formulae and much of the material is presented in terse lecture note format.

Handbook of Naturally Occurring Compounds (2 vols, T. K. Devon and A. I. Scott, Academic Press, 1972–) and *A Handbook of Alkaloids and Alkaloid-Containing Plants* (R. F. Raffauf, Wiley, 1970). These two handbooks contain data of interest to the natural product chemist; names, formulae and literature references are given.

The Alkaloids (R.H.F. Manske and H.L. Holmes, Eds, Academic Press,

Vols 1–4, 6–16, 1950–54, 1960–77; R.H.F. Manske, Ed., Vol. 5, 1955). This monograph, initially intended to be complete in five volumes, gives a comprehensive survey of the field of alkaloid chemistry and pharmacology. The occurrence, separation and properties of the more important alkaloids are covered in addition to their structure, stereochemistry, synthesis and pharmacology. Volume 1 contains a chapter 'Alkaloids in the plant', and Vol. 5 is heavily biased towards the pharmacology of the clinically important alkaloids. Volume 6 (1960) is a supplement to Vols 1 and 2, bringing them up to date; Vol. 7 is a supplement to Vols 2–5; while Vol. 8 (1965) covers the indole alkaloids. The practice of grouping chapters on similar alkaloids, either chemically or botanically, was abandoned after Vol. 8. Volumes 9–16 contain reviews of recent developments.

Ergebnisse der Alkaloid-Chemie bis 1960 (H.-G. Boit, Akademie Verlag, Berlin, 1961). This useful volume, which reviews the advances in alkaloid chemistry between 1950 and 1960 under the general headings of synthesis, degradations and reactions, and biosynthesis, is written in the style of *Annual Reports of the Chemical Society* but with many more structural formulae. There are many references to the original literature, and the less common alkaloids are described in tables under the headings of melting point, refractive index, derivatives, sources and functional groups.

Steroid Reactions, An Outline for Organic Chemists (C. Djerassi, Ed., Holden-Day, 1963). Although all the examples and literature references are for steroids, this book is virtually an abstract of functional group chemistry. The structural formulae, reaction conditions (e.g. 70% AcOH, 100°, 40 min) and the yield are given, and all reactions of the same type are grouped together. Some of the chapter headings are: Protection of carbonyl and alcohol groups; Introduction of double bonds; and Metal ammonia reductions of steroidal enones.

The Proteins (H. Neurath, Ed., 2nd edn, Academic Press, Vol. 1, 1963, Vol. 5, 1970). This is a monograph on protein chemistry which was originally intended to be complete in five volumes, although a 3rd edition is now in progress which is devoted to reviews of specific subjects. In the 2nd edition the field of protein composition, structure and function is comprehensively covered under such headings as: fractionation of proteins, amino acid analysis of polypeptides, X-ray analysis and protein structure, genetic control of protein structure, interaction of proteins with radiation (analytical), and structure and function of antigen and antibody proteins. Volume 5 is devoted to metalloproteins. The 3rd edition is edited by H. Neurath and R.L. Hill; Vol. 1 appeared in 1975 and Vol. 4 in 1978.

Medicinal Chemistry (A. Burger, Ed., 3rd edn, Wiley-Interscience, 1970). This work, in two parts, contains a collection of monographs on a wide range of synthetic drugs and their medicinal uses. Part I deals with general theories and presents drugs used in combating infections and parasitic diseases, as well as drugs applicable to neoplastic and immunochemical disorders. Part II contains chapters on biocatalysts and their structural variations, and also on drugs for the treatment of functional disorders.

Handbook of Solvents (I. Mellan, Reinhold, 1957–59). Volume 1 of this three-volume work is called *Handbook of Solvents* and deals with pure hydrocarbons; Vols 2 and 3 are called *Source Book of Industrial Solvents* and deal with halogenated hydrocarbons and with monohydric alcohols. The work has been written with the large-scale user in mind but will also be of use to the research worker.

Industrial Solvents Handbook (I. Mellan, Noyes Data Corp., 1970). The tables in this book contain pertinent data concerning the physical properties and degrees of solubility of solutes, etc. Much of the data is presented graphically and a number of phase diagrams are included. The text is organized according to solvent functional group.

SOCMA Handbook, Commercial Organic Chemical Names (Chemical Abstracts Service, 1967). This handbook was produced jointly by the Chemical Abstracts Service and the Synthetic Organic Chemical Manufacturers Association. It provides a means by which the chemical composition or structure of 6300 industrial organic compounds can be identified from a variety of their trade names. There are sections dealing with pure compounds and also with mixtures and polymers.

Elsevier Monographs on Toxic Agents (E. Browning, Ed., Elsevier, 1959–64). Several of the volumes in this series, edited by a former HM Medical Inspector of Factories, deal with organic compounds – e.g. *Toxic Aliphatic Fluorine Compounds* (F. L. M. Partison, 1959), *Carcinogenic and Chronic Toxic Hazards of Aromatic Amines* (T. S. Scott, 1962). Also useful is another work, *Toxicity and Metabolism of Industrial Solvents* (E. Browning, Elsevier, 1965), which contains data for over 150 compounds with original literature references.

NAMED REACTIONS

The practice of naming reactions after the discoverer and not on systematic grounds has become widespread in organic chemistry. This system gives a quick code which is very convenient for specialists in the narrow field but is inconvenient for others searching the literature for a particular reaction. Several books list these named reactions in alphabetical order and give information about the reaction with references to the literature. A useful list is given in the *Merck Index* (page 106) and details of three of the most useful collective works follow.

Name Index of Organic Reactions (J. E. Gowan and T. S. Wheeler, 2nd edn, Longmans, Green, 1960). Seven hundred and thirty-nine reactions are listed with a brief description, equation and leading references. The book contains a Type of Reaction Index in addition to a General Index of Reactants, Reagents and Products.

Name Reactions in Organic Chemistry (A. R. Surrey, 2nd edn, Academic Press, 1961). This book gives an autobiographical note on each named chemist in addition to the more usual information on the scope and limitations of the reaction, formulae and list of references. Over 125 reactions are discussed.

Organic Name Reactions (H. Krauch and W. Kunz, Wiley, 1964). Mechanistic details, yield and rate of reaction are given in this book in addition to the usual description, formulae and list of references for 523 named reactions.

SERIAL PUBLICATIONS

Encyclopaedic publications having been considered in the section on reference works, the serial works considered here are mainly of the review type which appear at fairly regular intervals. These reviews are written by specialists in the particular field and are therefore authoritative. Extensive bibliographies are a valuable feature of these publications, which are therefore useful in literature searches. The following list is a representative selection. (See Chapter 13 for the *High Polymers* series, etc.)

American Chemical Society, Monograph Series (Reinhold, 1921–71; American Chemical Society, 1972–   ). This well-established series of monographs on a variety of topics in pure and applied chemistry is most conveniently considered here; it incorporates a number of volumes of interest to the organic chemist. Typical titles include:

No. 147 *Organometallic Chemistry* (H. Zeiss, Ed., 1960)
No. 150 *Industrial Organic Nitrogen Compounds* (A. J. Astle, 1961)
No. 159 *Formaldehyde* (J. F. Walker, 3rd edn, 1964)
No. 170 *Regulation of Purine Biosynthesis* (J. F. Henderson, 1972)
No. 173 *Chemical Carcinogens* (C. E. Searle, 1976)

Chemical Society Specialist Periodical Reports (Chemical Society, 1967–). These volumes contain systematic and comprehensive review coverage of the progress of research in specific topics. The most widely expanding subjects are covered annually, while others appear biennially or less frequently. Certain titles, such as *Aromatic and Heteroaromatic Chemistry* and *Terpenes and Steroids*, are of importance to the organic chemist. Further details are given in Chapter 7.

MTP International Review of Science, Organic Chemistry (D. H. Hey, Ed., Butterworths, 1972–76). Ten volumes in the First Series of biennial reviews on organic chemistry, which covers literature published in 1970 and 1971, appeared in 1973. Each volume does not contain an index, these being collected in an additional index volume. Volume titles are: 1, *Structure Determination in Organic Chemistry*; 2, *Aliphatic Compounds*; 3, *Aromatic Compounds*; 4, *Heterocyclic Compounds*; 5, *Alicyclic Compounds*; 6, *Amino Acids, Peptides and Related Compounds*; 7. *Carbohydrates*; 8, *Steroids*; 9, *Alkaloids*; 10, *Free Radical Reactions*. The second series, covering literature published in 1972 and 1973, appeared in 1975–76. A Biochemistry Series has also been produced which could also prove of interest to the organic chemist. This series was discontinued in 1976.

Progress in Organic Chemistry (various editors, Butterworths, 1952–). Each volume in this series contains a number of specialist articles on recent developments in theoretical and experimental organic chemistry. A wide range of topics is covered and natural products are well represented. Surveys on Salamander alkaloids, Acidic hydrocarbons, Synthesis of prostaglandins and Chemiluminescence of organic compounds have appeared in recent volumes.

Organic Reactions (various editors, Wiley, 1942–). These volumes, which appear at roughly yearly intervals, contain review chapters each devoted to a single reaction of wide applicability. Experimental conditions for examples of each reaction are given in addition to a survey of the scope and limitations. Volume 23 appeared in 1976 and included chapters on Reduction and related reactions of α, β-unsaturated compounds with metals in liquid ammonia, Alkenes from tosylhydrazone and the Acyloin condensation.

Annual Reports in Organic Synthesis (various editors, Academic Press, 1970–   ) is an organized annual review of information of use to the synthetic organic chemist, compiled by abstracting about 40 primary chemistry journals, a list of which is given. There is no index; instead an extensive table of contents is provided. The information is presented in 'the form of structural equations giving details of yield, applicability and original literature reference.

Advances in Physical Organic Chemistry (V. Gold, Ed., Academic Press, 1963–   ). This series of fairly specialized review articles is directed mainly at the research worker or potential research worker in the field. The areas covered include both the theoretical and practical aspects, as shown by two recent chapter titles: Charge density – NMR chemical shift correlations in organic ions; Nucleophilic aromatic photosubstitution. *Progress in Physical Organic Chemistry* (A. Streitweiser and R. W. Taft *et al.*, Eds, Interscience, 1963–   ) provides more general reviews than those in the *Advances in Physical Organic Chemistry* series and should be of value to advanced students as well as non-specialists. Each chapter contains much tabulated numerical data.

Progress in Stereochemistry (W. Klyne and P. B. de la Mare, Eds, Vols 1–3; B. J. Aylett and M. M. Harris, Eds, Butterworths, 1969). Four volumes in this series have now appeared. They include reviews on the shapes of organic and inorganic molecules of varying complexity, together with chapters indicating the correlation between the stereochemistry and the chemical, physical and physiological properties of organic compounds.

Medicinal Chemistry (various editors, Wiley, 1951–   ) contains reviews on the pharmacological action of organic compounds. Systematic summaries of available data on the biological properties of substances studied are collected in tabular form at the end of each review.

Advances in Lipid Research (R. Paoletti and D. Kritchevsky, Eds, Academic Press, 1963–   ). The reports on subjects, including structural investigations, separations, metabolism and biosynthesis of lipids, contained in this series are designed mainly for research workers and specialists in the field. *Progress in the Chemistry of Fats and Other Lipids* (R. T. Holman, W. O. Lundberg and T. Malkin, Eds, 1952–63; R. T. Holman, Ed., Pergamon Press, 1963–   ) is a series of reviews and

monographs on the various aspects of lipid chemistry, including the
reactions, syntheses or analyses of different classes of lipid.

Advances in Protein Chemistry (various editors, Academic Press,
1944–   ) has appeared almost annually since 1951 and contains reviews on topics of current interest in protein chemistry. Reviews have
appeared on topics such as single proteins, groups of proteins, protein
structure and experimental techniques in protein chemistry and are
illustrated by the following recent examples: Phosphoproteins (1974);
Molecular orbital calculations on the conformation of amino acid
residues of proteins (1974); Membrane receptors and hormone action
(1976).

*Fortschritte der Chemie organischer Naturstoffe/Progress in the
Chemistry of Organic Natural Products* (founded by L. Zechmeister,
various editors, Springer, 1938–   ) is a valuable series of collected
reviews covering the synthesis, reactions, stereochemistry and properties of a wide range of natural products, in addition to the use of
physical methods in structure determination. The articles are contributed in English, French and German. Volume 34 appeared in 1977.
A cumulative index for Vols 1–20 (1938–62) appeared in 1964.

Advances in Carbohydrate Chemistry (various editors, Academic Press,
1945–68); *Advances in Carbohydrate Chemistry and Biochemistry*
(various editors, Academic Press, 1969–   ). This presents a series of
critical reviews on developments in carbohydrate chemistry covering
biochemical, industrial and analytical aspects. The original title was
altered after Vol. 23 to the new title (commencing as Vol. 24) to
emphasize the coverage of biochemical topics. A recent innovation has
been the inclusion of obituaries of prominent carbohydrate chemists.

Advances in Heterocyclic Chemistry (A. R. Katritzky, Ed., 1963–66;
A. R. Katritzky and A. J. Boulton, Eds, Academic Press, 1966–   )
covers a large and rapidly expanding field. Each volume from Vol. 12
(1970) contains a contents list for the earlier volumes. Supplement 1,
entitled *The Tautomerism of Heterocycles*, also appeared in 1976, updating the earlier reviews of this subject which originally appeared in
Vols 1 and 2 (1963).

Physical Methods in Heterocyclic Chemistry (A. R. Katritzky, Ed.,
Academic Press, 1963–   ) was originally planned to be complete in two
volumes covering non-spectroscopic and spectroscopic methods, respectively, which were published in 1963. However, the extensive growth of
the subjects caused the series to be recommenced. Volumes 3 and 4

(1971) updated the previous volumes, while Vol. 5 (1972) was devoted to X-ray structure analysis and Vol. 6 (1974) to additional new techniques.

Advances in Alicyclic Chemistry (H. Hart and G. J. Karabatsos, Eds, Academic Press, 1966–) is a collection of specialist monographs in the field of alicyclic chemistry.

Advances in Organometallic Chemistry (F. G. A. Stone and R. West, Eds, Academic Press, 1964–) contains authoritative reviews devoted to the complete field of organometallic chemistry; coverage also extends to the chemistry of complexes containing organic ligands. *Organometallic Reactions* (E. I. Becker and M. Tsutsui, Eds, Wiley, 1970–) is mainly concerned with the synthesis and reactions of selected categories of organometallic compounds. Many chapters contain tabulated examples using the same format as used in *Organic Reactions*.

Advances in Free Radical Chemistry (G. H. Williams, Ed., Logos Press, 1965–72; Elek, 1975–). This series, similar in design to the other *Advances* series, is devoted to the field of free radical chemistry.

Advances in Pest Control Research (R. L. Metcalf, Ed., Interscience, 1957–). Although designed mainly for the biologist, this series does contain some reviews of interest to organic chemists.

LABORATORY TECHNIQUES AND METHODS (INCLUDING ANALYSIS)

Advances in Organic Chemistry: Methods and Results (various editors, Interscience, 1960–) is a collection of review articles covering new techniques and preparative methods which are considered likely to be of wide application in the field of organic chemistry. Expertly written reviews cover the mechanism, scope and application of the reactions, in addition to the experimental details. Volume 7 (1970) was entirely devoted to the chemistry of cyclic enaminonitriles and *o*-aminonitriles.

Methoden der organischen Chemie, Houben-Weyl (E. Müller, Ed., 4th edn, Georg Thieme Verlag, 1952–). The new edition of this work, which is in German, deals comprehensively and critically with methods of all types, providing full experimental details for appropriate examples, in addition to the theoretical background. In this it follows closely the guiding principles laid down for the 3rd edition but has been considerably expanded to deal with the new material available and to give, for the first time, an extensive coverage of the patent literature. The new edition is planned to be complete in fifteen volumes, with a complete index published as Vol. 16, but, for technical reasons, the editors are publishing each volume or part as it becomes available and not in strict sequence. This, fortunately, is not too great an inconvenience, as each part is virtually an independent monograph. The work when completed will contain sections on analytical methods and the physical and chemical investigation of structure in addition to syntheses. Fields such as heterocyclic chemistry, carbohydrates, peptides and proteins and polymerization will be covered for the first time in the 4th edition.

Newer Methods of Preparative Organic Chemistry (W. Foerst, Ed., Vol. 1, Interscience, 1948; Vols 2– , Academic Press, 1963–). The reviews in Vol. 1 give details of the scope and limitations, preparation of reagents and experimental data on a number of important reactions and reagents. It is a selection from a collection of articles published in Germany in 1944 and this translation was published in 1948 by consent of the Alien Property Custodian. Volumes 2–6 are translations of collections of reviews which originally appeared in *Angewandte Chemie*. These reviews have been considerably expanded and numerous experimental procedures have also been included.

Organic Syntheses (various editors, Wiley, 1921–). This annual series gives satisfactory methods for the synthesis of individual organic compounds and each method is checked in two independent laboratories before publication. Volumes 1–30 are now out of print but five revised collective volumes have now been issued: I (Vols 1–9, 2nd edn, 1941); II (Vols 10–19, 1943); III (Vols 20–29, 1955); IV (Vols 30–39, 1963); V (Vols 40–49, 1973). A useful feature of all volumes is the very comprehensive indexing. A Cumulative Index for Collective Volumes I–V was published in 1976. Volume 50 (1970) and all subsequent volumes contain brief references to unchecked procedures received for publication, giving equations and yields. Full details of these procedures, whether subsequently accepted or rejected for publication, may be obtained from the publishers.

Synthetic Methods of Organic Chemistry (W. Theilheimer, Ed., Karger, Basle, 1948–). An English edition is available of this work, which abstracts new methods for the synthesis of organic compounds. The abstracts appraise the applicability of the method, show the number of steps and the yield and give references to the original paper or to a usable abstract for experimental details. The syntheses are arranged systematically in each volume by reaction type and author names are not used. The indexes include a name index, a reaction type index, a reagent list and a list of supplementary references. Volume 30 (1976) contains a cumulative index to Vols 26–30; the next cumulative index will appear in Vol. 35.

Technique of Organic Chemistry (A. Weissberger, Ed., Interscience, 1948–63). This thorough and comprehensive treatise in fourteen volumes gives a critical survey of techniques used in the organic laboratory, placing emphasis on theoretical background throughout but in no way neglecting reference to practical details. Topics covered include physical methods of organic chemistry, separation and purification, distillation, organic solvents, and the various techniques of chromatography and spectroscopy. Volume 1 is in the 3rd edition and Vols 2–4 and 7–9 are in the 2nd edition.

Techniques of Chemistry (A. Weissberger, Ed., Interscience, 1971–). This treatise is the successor to the *Technique of Organic Chemistry* series, formed by combination with the companion series *Technique of Inorganic Chemistry*. The new series contains later editions of works of the previous series including: Vol. 1, *Physical Methods of Chemistry* (4th edn, previously Vol. 1 of *T.O.C.*); Vol. 2, *Organic Solvents* (3rd edn, previously Vol. 7 of *T.O.C.*); Vol. 4, *Elucidation of Structures by Physical and Chemical Techniques* (2nd edn, previously Vol. 11 of *T.O.C.*); Vol. 6, *Investigation of Rates and Mechanisms of Reactions* (3rd edn, previously Vol. 8 of *T.O.C.*).

Organic Synthesis (2 vols, V. Migrdichian, Reinhold, 1957) is a summary of the preparation and reactions of organic functional groups which gives a good survey of the older literature. *Synthetic Organic Chemistry* (R. B. Wagner and H. D. Zook, Wiley, 1953). A single volume which summarizes the most common syntheses of organic mono- and difunctional compounds. Very few experimental details are given but there are over 6000 references to original papers which contain explicit details. Each chapter is devoted to an individual functional group and many of the relevant compounds are listed in tables giving yield, m.p./b.p. and references.

Survey of Organic Syntheses (C. A. Buehler and D. E. Pearson, Wiley, 1970) lists the principal methods for the synthesis of the main types of organic compounds. Each chapter is devoted to the creation of one functional group from another. The equation for each reaction is given and its value and limitations discussed, together with basic theory.

Specific examples are also given in sufficient detail to permit laboratory preparation.

Organic Synthesis with Isotopes (2 parts, A. Murray and D. L. Williams, Interscience, 1958). Many new methods have been devised for the synthesis of labelled compounds in order to increase the yield or to obtain maximum utilization of the isotopic reagent; this work describes many of these methods in sufficient detail to be used in the laboratory. Part 1 deals with compounds of isotopic carbon, and Part 2 with compounds labelled with halogens, hydrogen, nitrogen, oxygen, phosphorus and sulphur.

Purification of Laboratory Chemicals (D. D. Perrin, W. L. F. Armarego and D. R. Perrin, Pergamon Press, 1966) is in three sections: the first is a discussion, with theoretical background, of general purification techniques; the second outlines the purification method for each of a large number of organic compounds; and the third discusses general principles of purification for use with compounds not listed in the second section.

Reagents for Organic Synthesis (L. F. and M. Fieser, Wiley, 1967—   ). Volume 1 contains the structural formula, molecular weight, physical constants, preparation, purification, commercial suppliers and applications for 1120 reagents of use to the organic chemist. These are arranged in alphabetical order and literature references are given, the coverage being up to early 1966. Supplementary volumes 2—5 (1975) cover recent developments on previous reagents, with cross-references to the data in earlier volumes and also new reagents. These later volumes do not cumulate.

Synthetic Reagents (2 vols, J. S. Pizey, Ellis Horwood, 1974). These volumes give thorough coverage of the use of a selected group of very versatile reagents, and complement the work of Fieser and Fieser. Among the reagents treated are: lithium aluminium hydride and thionyl chloride (Vol. 1), and diazomethane and Raney nickel (Vol. 2).

Compendium of Organic Synthetic Methods (I. T. and S. Harrison, Wiley, 1971—   ). The original volume (1971) is a systematic listing of functional group transformations designed for use by synthetic organic chemists. Reactions are classified on the basis of the functional group of starting material and product, and there is a very useful quick reference summary table. The information given is very terse: an equation, reaction conditions, catalyst, yield and abbreviated original literature reference(s). Volume 2 (1974) is a supplement to the original work, with a chapter on difunctional compounds. Both volumes also cover methods for the protection of functional groups.

Free Radical Reactions in Preparative Organic Chemistry (G. Sosnovsky, Macmillan, 1964) is arranged in sections which cover various types of compound and lists the reaction conditions with complete information on temperature, pressure, catalyst, initiator and solvents.

Organometallic Compounds, Methods of Synthesis, Physical Constants and Chemical Reactions (M. Dub, Ed., 2nd edn, Springer, 1966–   ) is in three volumes: Vol. 1 contains derivatives of the transition metals of Groups III to VIII of the periodic table; Vol. 2 contains Ge, Sn and Pb; while Vol. 3 is devoted to As, Sb and Bi. The 2nd edition covers the literature from 1937 to 1964 and provides a comprehensive, non-critical source of information on organometallic compounds; a formula index is also available. The first supplements to Vols 1–3 scanning the literature from 1965 to 1968 appeared in 1975, 1973 and 1972, respectively.

Methods in Carbohydrate Chemistry (R. L. Whistler *et al.*, Eds, Academic Press, 1962–   ). This series contains descriptions of reliable methods for the handling, preparation, reaction and analysis of carbohydrates, and the field covered ranges from simple monosaccharides to complex polysaccharides.

Industrial Chemicals (F. A. Lowenheim and M. K. Moran, 4th edn, Wiley, 1975) is a revision of the earlier work of Faith, Keyes and Clark. The manufacture (with flow diagram and details), properties, uses, economics, transportation requirements and plant sites are given for a large number of important industrial chemicals, including many organic compounds.

A Laboratory Manual of Analytical Methods in Protein Chemistry (including Polypeptides) (P. Alexander *et al.*, Eds, Pergamon Press, 1960–). Volumes 1–3 (1960–61) contained reviews on the manipulation of proteins – e.g. their Separation and isolation (Vol. 1); Composition, structure and reactivity (Vol. 2); and Determination of size and shape of protein molecules (Vol. 3). The later volumes contain reviews of recent developments and new techniques.

Tables of Resolving Agents and Optical Resolutions (S. H. Wilen, University of Notre Dame Press, 1972) is a collation of resolving agents and resolutions. It is tabulated and contains many references to the original literature.

Organic Analysis (J. Mitchell, I. M. Kolthoff, E. S. Proskauer and A. Weissberger, Eds, Interscience, Vol. 1, 1953; Vol. 4, 1960) is a series which aims to present up-to-date information on quantitative chemical and physical methods of organic analysis. The first three volumes are concerned with functional group analysis, and the fourth gives surveys of newer instrumental techniques and includes a chapter on enzymic analytical reactions. Other still useful works are:

Organic Functional Group Analysis by Micro and Semimicro Methods (N. D. Cheronis and T. S. Ma, Interscience, 1964)
Quantitative Organic Microanalysis (A. Steyermark, 2nd edn, Academic Press, 1961)
Spot Tests in Organic Analysis (F. Feigl, 7th edn, Elsevier, 1966)

Elementary Analysis Tables (G. Ege, Verlag Chemie/Wiley, 1966). These computer-prepared anti-composition tables give a range of possible empirical formulae in the range C_1-C_{40} for any combination of elemental analysis figures for compounds containing carbon, hydrogen, nitrogen or oxygen (sulphur). *Tables of Percentage Composition of Organic Compounds* (H. Gysel, 2nd edn, Birkhaüser, 1969) presents tables of elemental analyses and molecular weights for a wide range of empirical formulae containing C, H, N and O (S). Formulae C_1-C_{50} (hydrocarbons to C_{60}, and C, H, O to C_{52}) are included. *Computer Compilation of Molecular Weights and Percentage Composition for Organic Compounds* (M.J.S. Dewar and R. Jones, Pergamon Press, 1969) gives tables of elemental analyses, molecular weights and isomolecular weights (for use in mass spectrometry) for a wide range of empirical formulae containing C, H, and one or two heteroatoms (heteroatoms are Br, Cl, F, I, N, O, P). Formulae C_1-C_{30} (hydrocarbons to C_{40}) are included. The instructions are written in English, French and German in the above three works.

Melting Point Tables of Organic Compounds (W. Utermark and W. Schicke, 2nd edn, Interscience, 1963) gives the name, structure, empirical formula, molecular weight, boiling point, refractive index, *Beilstein* reference and some properties and reactions of a large number of organic compounds arranged in order of melting point from $-189.9°$ to $500°C$. *Handbook of Tables for Organic Compound Identification* (Z. Rappoport, 3rd edn, Chemical Rubber Co., 1967) is a supplement to the *Handbook of Chemistry and Physics* which includes the melting or boiling points and, in some cases, the refractive index and/or densities of over 8150 organic compounds, in addition to the melting points of some of their more common derivatives. (See also Chapter 8.)

Chromatographic and Electrophoretic Techniques (2 vols, I. Smith, 3rd edn, Heinemann, 1968–69). The bias in these two volumes is very much towards the practical aspects of chromatography (Vol. 1) and electrophoresis (Vol. 2), and they are written by a man with wide experience of these fields.

Separation Methods in Biochemistry (C. J. O. R. and P. Morris, 2nd edn, Pitman, 1976) is as useful to organic chemists as it is to biochemists, but the approach is more theoretical than practical.

Bibliography of Paper Chromatography and Survey of Applications (I. M. Hais and K. Macek, Eds, Czechoslovak Academy of Sciences/ Academic Press, Vol. 1, 1960; Vol. 2, 1962). These two volumes list titles of original papers, authors and references to the journal, year and page. The entries are grouped by compounds separated, and within each group they are listed under authors in alphabetical order. Volume 1 covers the literature from 1944 to 1956; Vol. 2 from 1957 to 1960. Volume 1 is published in Czech but a supplement is provided to enable English-speaking readers to use it. The title and coverage have been extended to become *Bibliography of Paper and Thin-Layer Chromatography* (K. Macek *et al.*, Eds, Elsevier), published as Supplementary Volumes to the *Journal of Chromatography*.

Suppl. Vol. 1 (1968) covers 1961–65; Suppl. Vol. 2 (1972) covers 1966–69; Suppl. Vol. 5 (1976) covers 1970–73. Other Supplementary Volumes in these series survey the following fields: Suppl. Vol. 3 (1973), Column chromatography, 1967–70; Suppl. Vol. 4 (1975), Electrophoresis, 1968–72.

Gas Chromatography (A. B. Littlewood, 2nd edn, Academic Press, 1970) is a useful general text which is designed as a reference work for people using this important technique. The large theoretical section contains chapters on the theory of the design and operation of columns and on the principles of detection. The practical section describes, in general terms, the separation of compounds by classes. References are given to papers containing the practical details. *Practical Manual of Gas Chromatography* (J. Tranchant, Ed., Elsevier, 1969) was written to assist with the practical problems of this analytical technique. A useful bibliography is given in the Foreword.

Thin Layer Chromatography, a Laboratory Handbook (E. Stahl, Ed., 2nd edn, Allen and Unwin/Springer, 1969) is a useful handbook dealing with instrumentation, techniques and applications of thin layer chromatography to a wide range of organic compounds. Much tabulated data and a multitude of literature references are included.

Solvents Guide (C. Marsden and S. Mann, 2nd edn, Cleaver-Hume,

1963) lists the sources, physiological properties and storing and handling techniques for a wide range of solvents and also gives information
about manufacturers.

PHYSICAL METHODS OF STRUCTURE DETERMINATION

Physical methods have made such an enormous contribution to structural determination in organic chemistry in recent years that no account
of organic chemistry would be complete without mentioning them. A
large number of tables and books have appeared in the last few years to
satisfy the demand for information in this rapidly expanding field. The
following list is meant to supplement with interpretative works the
section on spectroscopic data collections dealt with in Chapter 8, and
includes examples from the most important fields, including a section
on nuclear magnetic resonance, which was not covered in the previous
edition. A useful general introduction to many methods may also be
found in many organic chemistry textbooks, and in Weissberger,
Houben-Weyl, etc.

General texts

Determination of Organic Structures by Physical Methods (F. C. Nachod
et al., Eds, Academic Press, Vol. 1, 1955; Vol. 6, 1976). In the original
work (1955) a chapter is devoted to each physical method and technique; interpretation of results and applications are discussed. The
methods covered include optical rotation, magnetic susceptibilities,
dipole moments and X-ray diffraction, in addition to the more usual
IR, MS, NMR, UV and visible spectroscopy. Later volumes contain
expanded coverage of the techniques; Vol. 4 is entirely devoted to the
various techniques of NMR, including ^{15}N, ^{13}C and ^{31}P nuclei.

Applications of Spectroscopy to Organic Chemistry (J. C. D. Brand and
G. Eglinton, Oldbourne, 1965) and *Spectroscopic Methods in Organic
Chemistry* (D. H. Williams and I. Fleming, 2nd edn, McGraw-Hill, 1973)
are two very useful basic texts which cover the techniques of UV,
visible, IR and NMR spectroscopy. Williams and Fleming also covers
MS.

Physical Methods in Heterocyclic Chemistry (A. R. Katritzky, Ed.,
Academic Press, 1963–74) is a comprehensive account of the applica-
tions of physical methods to the wide field of heterocyclic chemistry.
Volume 1 covered non-spectroscopic methods and Vol. 2 covered

spectroscopic methods; these both appeared in 1963. The series was updated later, Vols 3 and 4 (1971) covering methods other than X-ray structural analysis, while Vol. 5 (1972), sub-titled 'Handbook of Molecular Dimensions', comprised tabulated data for 1367 compounds. Volume 6 (1974) was devoted to the more recently introduced techniques, including UV, photoelectron spectra and microwave spectroscopy. The emphasis of the series is mainly upon the interpretation of data.

Handbook of Organic Structural Analysis (Y. Yukawa, Ed., Benjamin, 1965). This collection comprises 20 000 entries covering a wide range of physical properties. Dissociation constants, IR, NMR, optical rotatory dispersion, redox potentials, thermochemical constants and UV are among the range of physical properties listed. Within each section active groups (e.g. chromophores for spectroscopy, acidic or basic groups for dissociation constants) are listed with the relevant physical data and literature references for a number of model compounds.

Infra-red

The Infra-red Spectra of Complex Molecules Vol. 1 (L. J. Bellamy, 3rd edn, Chapman and Hall, 1975); Vol. 2 *Advances in Infrared Group Frequencies* (L. J. Bellamy, Methuen,* 1968). Dr. Bellamy's original book (1950; 2nd edn, 1958) is generally accepted as the standard text in this field. It presents a critical review of the data on which the usual empirical spectra/structure correlation tables are based. The book contains correlation tables, an empirical discussion of the characteristic bands for the more important classes of compound and the factors which can influence their frequencies. Volume 2 (1968) is a supplement to the earlier book. Volume 1 essentially presents the data, while Vol. 2 discusses the reasons for these known facts in chemical and physical terms.

An Introduction to Practical Infra-red Spectroscopy (A. D. Cross and R. A. Jones, 3rd edn, Butterworths, 1969) and *Characteristic Frequencies of Chemical Groups in the Infra Red* (M. St C. Flett, Elsevier, 1963) are two excellent texts containing correlation tables and short discussions on each of the more common functional groups. They are particularly suitable for undergraduates or for occasional users dealing with fairly straightforward compounds.

---

*Now available through Chapman and Hall.

Laboratory Methods in Infrared Spectroscopy (R. G. J. Miller and B. C. Stace, Eds, Heyden, 1972) concentrates upon the practical problems involved in obtaining an infra-red spectrum. It is very useful for the spectroscopist with a problem.

Mass spectrometry

Applications of Mass Spectrometry to Organic Chemistry (R. I. Reed, Academic Press, 1966) is a useful introductory text which contains chapters on the production of spectra and on the fragmentation patterns of compounds of different groups. It contains many tables of examples but only a few actual spectra.

Mass Spectrometry of Organic Compounds (H. Budzikiewicz, C. Djerrasi and D. H. Williams, Holden-Day, 1967); *Structural Elucidation of Natural Products by Mass Spectrometry*. Vol. 1, *Alkaloids*; Vol. 2, *Steroids, Terpenoids and Sugars* (same authors, Holden-Day, 1964). These three related volumes give fragmentation patterns for selected compounds and groups of compounds with explanations and specimen spectra. A good companion text to these is *The Mass Spectra of Organic Molecules* (J. H. Beynon, R. A. Saunders and A. E. Williams, Elsevier, 1968). A complementary work is *Biochemical Applications of Mass Spectrometry* (G. R. Walker, Ed., Wiley, 1972), which discusses the fragmentation patterns of important biochemical molecules.

Mass and Abundance Tables for Use in Mass Spectrometry (J. H. Beynon and A. E. Williams, Elsevier, 1963), *Table of Meta-Stable Transitions for Use in Mass Spectrometry* (J. H. Beynon, R. A. Saunders and A. E. Williams, Elsevier, 1965) and *Table of Ion Energies for Metastable Transitions in Mass Spectrometry* (J. H. Beynon, R. M. Caprioli, A. W. Kunderd and R. B. Spencer, Elsevier, 1970). These sets of tables (see also Chapter 8) are computer-calculated and printed with instructions in several languages. They assist in the extraction of data from mass spectra and are of considerable assistance in their interpretation. The third book is designed to aid in the interpretation of ion kinetic energy spectra.

Tables for Use in High Resolution Mass Spectrometry (R. Binks, J. S. Littler and R. L. Cleaver, Heyden/Sadtler, 1970). These tables are useful to the operators of mass spectrometers for the technique of mass matching, as used to determine accurate masses and, hence, elemental compositions.

Ultra-violet

The Theory of the Electronic Spectra of Organic Molecules (J. N. Murrell, Chapman and Hall, 1971) is a reprint of the original volume published by Methuen in 1963 and presents a quantum mechanical approach to the interpretation of spectra, with a basic introduction to the necessary quantum mechanics.

Gillam and Stern's Introduction to Electronic Absorption Spectroscopy in Organic Chemistry (E. S. Stern and C. J. Timmons, 3rd edn, Arnold, 1970) is a revised edition of the original work, which previously appeared in 1958. It is a more empirical and practical book, containing an introduction to light absorption, the origin of spectra and some experimental techniques, in addition to a discussion of the absorption data for typical chromophores and an outline of some applications in analytical and structure determination work.

Nuclear magnetic resonance

General/proton

High Resolution Nuclear Magnetic Resonance Spectroscopy (2 vols, J. W. Emsley, J. Feeney and L. H. Sutcliffe, Pergamon Press, 1965–66). Volume 1 (1965) of this comprehensive treatise is concerned with basic theory and spectral analysis, while Vol. 2 (1966) deals with structural applications. The complete work contains a considerable quantity of data for reference purposes. Despite the very significant advances that have been made in the field of NMR since the publication of these volumes, they are nevertheless still of great value. A less comprehensive text is *Applications of Nuclear Magnetic Resonance Spectroscopy in Organic Chemistry* (L. M. Jackman and S. Sternhell, 2nd edn, Pergamon Press, 1969), but it provides a thorough basic theoretical background, necessary for the interpretation of NMR spectra. It, too, contains much tabulated data and many original literature references.

Interpretation of NMR Spectra (K. B. Wiberg and B. J. Nist, Benjamin, 1962) contains a collection of theoretically calculated second-order NMR spectra presented in a computer-type visual format. It is a very useful manual for the estimation of coupling constants and chemical shifts from complex spectra.

Other nuclei

Carbon-13 Nuclear Magnetic Resonance for Organic Chemists (G. C. Levy and G. L. Nelson, Wiley, 1972) is a useful introductory text to the recently introduced and rapidly expanding field of ^{13}C NMR.

Carbon-13 NMR Spectroscopy (J. B. Stothers, Vol. 24 of *Organic Chemistry*, a series of monographs; see above) and *^{13}C NMR Spectroscopy Methods and Applications* (E. Breitmaier and W. Voelter, Verlag Chemie, 1974) are later, more comprehensive texts which include much tabulated data and many literature references. They also include an introduction to the techniques of pulsed NMR and Fourier transformation.

Nitrogen NMR (M. Witanowski and G. A. Webb, Eds, Plenum Press, 1973) deals with the theory, technique and applications of ^{14}N and ^{15}N NMR.

PROOF ADDITIONS

Comprehensive Organic Chemistry (D. H. R. Barton and W. D. Ollis, Eds, 6 vols, Pergamon Press, 1979) is a medium-sized work designed to fill the gap between the smaller and less informative texts on organic chemistry and existing multi-volume treatises such as *Rodd* (page 172).

Tetrahedron Reports on Organic Chemistry (D. H. R. Barton, Ed., Pergamon Press, Vols 1–3, 1978). These volumes contain reviews reflecting the breadth of interest and endeavour of the modern organic chemist. Each volume comprises ten Reports previously published in *Tetrahedron*.

13

Polymer science

R. T. Bottle and *J. M. Sweeney*

Polymer science comes second only to the pharmacomedical sciences as a field where industrial interest in research and development is very high. Indeed, the Chemical Abstracts Service, with its eye on commercial possibilities, chose this field for its second specialist alerting service, *POST*, two years after introducing *CBAC* (q.v.). While the more technological and engineering aspects of polymers are excluded from this chapter, the literature frequently divides itself into works dealing with either fundamental or applied aspects of polymer science. On the fundamental side two distinct approaches are often discernible (usually connected with the authors' backgrounds). These are that of the organic chemist on the one hand and that of the physical chemist or physicist on the other. The polymers themselves can be classified according to whether they are natural products (chemically modified or not) or totally synthetic in origin. Although inorganic polymers are of increasing interest, the vast bulk of the literature deals with organic polymers.

The uses to which commercially important polymers are put depend mainly on whether their properties are those of plastics (thermoplastic or thermosetting), elastomers (natural or synthetic rubbers) or fibres. Apart from the materials which these divisions suggest, applied polymer science covers such materials as paints and coatings, paper, packaging, leather, adhesives, insulating foams, ion exchangers, etc., while natural polymers such as starch and the proteins are important foodstuffs. An interesting history which shows many of the uses of polymers is *The First Century of Plastics* (M. Kaufman, Plastics Institute, London, 1962), while to mark the centenary of parkesine, the first plastic, ICI Plastics Division published *Landmarks of the Plastics Industry* (1962).

194

H. Staudinger's *Arbeitserinnerungen* (Huthig, Heidelberg, 1961) is an interesting personal history by the Nobel Laureate who laid the foundations of modern polymer chemistry. It contains a bibliography of Staudinger's work and lists 150 dissertations originating from his school, and was translated as *From Organic Chemistry to Macromolecules: a Scientific Autobiography based on my original papers* (Wiley, 1970).

THE PRIMARY LITERATURE AND ITS EXPLOITATION

Journal literature is still the major form of primary literature, although much of the original information on, for example, the low-pressure polymerization of ethylene and propylene came piecemeal from patents. Indeed, in the years following the discovery of this process the volume of relevant patent literature exceeded that of journal literature by a ratio of about three to one. However, the amount of patent literature subsequently declined and from about twelve years after the initial discoveries the journal literature became the major contributor of the two (J.M. Sweeney, M.Sc. Thesis, The City University, 1976). The most important specialist journals are *Makromolekulare Chemie* and *Journal of Polymer Science*. The latter split into several parts in 1963. The parts now in issue are A1 *Polymer Chemistry*, A2 *Polymer Physics*, B *Polymer Letters* and C *Polymer Symposia*. *Biopolymers* was inaugurated at the time of the split. B is an important source of preliminary communications in this field. The symposia reported in C are predominantly American. Advance information on ACS Division of Polymer Chemistry symposia is obtained from *Polymer Preprints*, which may be obtained on subscription by libraries. (The papers are normally re-published in *Polymer Symposia* or in *Journal of Macromolecular Science* Part A without recording the usually valuable discussion.) Each issue contains useful information on the symposia planned for the next half-yearly meeting and notes other polymer meetings. Since 1965 Interscience have been publishing separately *Applied Polymer Symposia*, which complement the *Journal of Applied Polymer Science*. Papers presented at ACS meetings are now indexed separately by *POST* (q.v.). ACS also publish the important *Rubber Chemistry and Technology*.

Another journal to split into parts is *Journal of Macromolecular Science*, which in 1967 was divided into A *Chemistry*, B *Physics*, C *Reviews* (previously published as *Reviews in Macromolecular Chemistry*) and D *Polymer Processing and Technology* (reviews). In 1970 the latter became *Polymer-Plastics Technology and Engineering*.

196   *Polymer Science*

A new 'quickie' journal, *Polymer Bulletin*, was inaugurated in 1978 by Springer.

The important Russian journal *Vysokomolekulyarnye Soedineniya Seriya A* is translated cover-to-cover as *Polymer Science USSR. Seriya B (Kratkie soobshcheniya)*, which split off from *Seriya A* in 1967, is devoted to short communications.

One review journal has already been mentioned and another is *Advances in Polymer Science (Fortschritte der Hochpolymeren-Forschung)*, in which most of the reviews are in English. Review serials include *Progress in High Polymers, Progress in Polymer Science* and *Annual Report on the Progress of Rubber Technology*, while some reviews of interest will also be found in *Advances in Food Research, Modern Materials*, etc., as well as in *Advances in Protein Chemistry*, etc.

Apart from reviews, the primary literature may be monitored through a variety of abstracting services. The most important of these is *Chemical Abstracts*, which has been discussed in detail in Chapter 5 and which makes its *Macromolecular Sections* available for separate subscription. The machine-readable version is available as bi-weekly tapes and is called *Polymer Science and Technology (POST)*. About 50 000 abstracts per year are issued and patents from 26 countries are included. The subscription includes the right to distribute the results of searching the POST file within the subscriber's organization. POST was formerly divided into journal and patents sections which appeared alternately, but these are now combined.

Derwent Publications also produce a patents documentation and alerting service on plastics, rubbers, etc., known as *Plasdoc*. This is a section of *CPI*, which is discussed in Chapter 14. The complete service is expensive but a lot of documentation is provided, including Individual Country File Booklets, an Abstracts journal (containing a Patent Concordance), Microfilmed Basic Patents, Coded Material on Cards, Computer-generated Indexes, Tapes and Search Facilities. Derwent were considering issuing a service similar in coverage to *POST* to be known as *Polydoc*. This will not now appear owing to lack of support.

German patents are included in the monthly *Literatur-Schnelldienst, Kunststoffe Kautschuk Fasern* (Deutsches Kunststoff-Institut, Darmstadt, 1955–). The Rubber and Plastics Research Association (Shawbury, Shropshire) publishes *RAPRA Abstracts*, formed by the combination of *Plastics: RAPRA Abstracts* and *Rubbers: RAPRA Abstracts* in 1968. Each fortnightly issue contains abstracts drawn from the journal and patent literature. Sections on forthcoming conferences, book reviews and abstracts of articles of economic interest are all included. There are author, subject and patent indexes for each cumulated volume. RAPRA also publishes the monthly *International Polymer Science and Technology*, which contains abstracts of papers in Russian,

East European and Japanese journals. Subscribers can request full translations of these papers and several of these translations appear in each issue. A number of commercial firms and consultants publish abstracts in this field, including the monthly *Polymerics* (Yarsley Labs, Redhill, Surrey), and the weekly *Plastics Abstracts* (Plastics Investigations, Welwyn, Herts.), although the latter is confined to UK patents.

The exploitation of the primary literature of polymer science follows the same principles as those laid down in Chapter 18. Apart from short-cuts provided by the well-documented review, *Chemical Abstracts* is the searching tool of choice, although in the more technological areas *RAPRA Abstracts* is useful. *Science Citation Index* (Chapter 5) may also be of assistance, especially in the biopolymer field. *Plasdoc* informs on new patents (especially for European subscribers) rather more quickly than *Chemical Abstracts*. Some of the other services mentioned will prove useful where budgets will not run to *Chemical Abstracts* or *Plasdoc*.

FUNDAMENTAL STUDIES

Among the most prolific publishers in this field are Wiley-Interscience. They have issued a classic monograph series under the generic title *High Polymers* which has an impressive editorial board dominated by the doyen of American polymer chemists, H. Mark. The first volume, the collected papers of W. H. Carothers (the inventor of nylon), appeared in 1940. So far more than twenty-five volumes have appeared, several going to second editions. At least six of the early volumes are now out of print. This series is complemented by *Polymer Reviews* (Interscience, 1958–   ), under the general editorship of H. Mark and E. H. Immergut. Another Wiley publication, *Macromolecular Reviews* (A. Peterlin, Ed.), reached its tenth volume in 1976. Each volume contains several reviews on different aspects of polymer science.

Over 150 detailed recipes, which have been independently tested outside the authors' laboratories, have appeared in the first six volumes (1963–77) of Wiley-Interscience's series of *Macromolecular Syntheses*. Volumes 1–5 (J.A. Moore, Ed.) were cumulated in 1978. A more general text is *Preparative Methods of Polymer Chemistry* (W.R. Sorenson and T. W. Campbell, Interscience, 2nd edn, 1968), while various polymerization methods are described in *Polymerization Processes* (C. E. Schildknecht, Wiley, 1978, *HP XXIX*). *Monomers* (E. R. Blout, W. P. Hohenstein and H. Mark, Eds, Interscience, 1949–54) is a loose-leaf compilation of data on methods of preparation, purification and polymerization of monomers which contains some still useful information but has largely been superseded by *Polymer Handbook* (q.v.). There are some general works on polymerization – for example,

Reactivity, Mechanism and Structure in Polymer Chemistry (A. D. Jenkins and A. Ledwith, Eds, Wiley, 1974) and *Kinetics and Mechanisms of Polymerisation Reactions* (P. E. M. Allen and C. R. Patrick, Wiley, 1974), which is, however, concerned mainly with addition polymerization. Several volumes of the *High Polymers* and *Polymer Reviews* series deal with specific preparative aspects such as *Emulsion Polymerization* (F. A. Bovey, I. M. Kolthoff, A. J. Medalia and E. J. Meechan, 1966, *HP IX*) or *Condensation Polymers: by Interfacial and Solution Methods* (P. W. Morgan, 1965, *PR X*). A recent treatment of the former is *Emulsion Polymerisation. Theory and Practice* (D. C. Blackley, Applied Science, 1975). The important techniques of block and graft polymerization are dealt with in Vol. 4 of Plenum's series *Polymer Science and Technology*, namely *Recent Advances in Polymer Blends, Grafts and Blocks* (L. H. Sperling, Ed., 1974) and also in *Block and Graft Copolymerization* (2 vols, R. J. Ceresa, Ed., Wiley, 1973, 1976). Another polymerization process which has become very important in the last twenty years is the subject of *Kinetics of Ziegler–Natta Polymerization* (T. Keii, Chapman and Hall, 1972). Plastics, rubbers, fibres and other polymers are normally still capable of undergoing further reactions and these are discussed in *Reactions on Polymers* (J. A. Moore, Ed., Reidel, 1973). A parallel problem to preparation is that dealt with in *Polymer Fractionation* (M. J. R. Cantow, Ed., Academic Press, 1967). A modern introductory text dealing with all types of polymers is *Textbook of Polymer Science* (F. W. Billmeyer, 2nd edn, Wiley, 1971). *Experiments in Polymer Science* (E. A. Collins, J. Bares and F. W. Billmeyer, Wiley, 1973) is a practical manual intended as a companion volume. *Organic Polymer Chemistry* (K. J. Saunders, Chapman and Hall, 1973) surveys an important division of polymer chemistry.

Specific classes often form the subject of monographs either as part of a series or independently and the most important types are covered by at least one definitive work. Examples from the *High Polymers* series are *Fluoropolymers* (L. A. Wall, Ed., 1972, *HP XXV*) and *Allyl Compounds and their Polymers* (C. E. Schildknecht, 1973, *HP XXVIII*). The latter includes polyolefins and provides a comprehensive treatment of the subject. Natural polymers are also covered by books, as, for example, A. J. Radley's *Starch and its Derivatives* (4th edn, Chapman and Hall, 1968). H. J. Stern's *Rubber – Natural and Synthetic* (2nd edn, Maclaren, 1966) stresses the applied rather more than the fundamental aspects.

Polymers can be classed by particular characteristics as well as by chemical structure. Recent studies based on this approach are *The Physics of Glassy Polymers* (R. N. Howard, Ed., 1973), *Ionic Polymers*

(L. Holliday, Ed., 1975), *Toughened Plastics* (C. B. Bucknall, 1977) and *Structure and Properties of Oriented Polymers* (I. M. Ward, Ed., 1975). These four books are in the Applied Science *Materials Science Series*. Another work of this type is *The Stereo Rubbers* (W. M. Saltman, Ed., Wiley, 1977).

Information about specific polymers may also be obtained from the Kirk–Othmer *Encyclopedia of Chemical Technology* (see Chapter 7) or in rather more detail from the *Encyclopedia of Polymer Science and Technology* (15 vols plus index vol. and 2 supplement vols, H. F. Mark, N. G. Gaylord and N. Bikales, Eds, Interscience, 1964–). The latter is also useful for analytical and testing methods and each article is well documented. As in Kirk–Othmer, American terminology is used and recent American production and other statistics are given. *Kunststoff-Handbuch* (R. Vieweg, Ed., Carl Hanser, Munich, 1963–) is a twelve-volume German encyclopaedia. After the first general volume on processing, testing, etc., each of the remaining volumes deals with a specific type of polymer and its associated technology.

A polymer sample bank was established in 1967 at the Polymer Institute, 333 Jay Street, Brooklyn, NY 11201, as a non-profit organization to supply a stock of non-varying polymer samples. Dow Chemicals supply polystyrene latex dispersions of standardized narrow particle size range, though these are really of more interest to the colloid scientist.

The converse of polymer synthesis is discussed in *Degradation and Stabilisation of Polymers* (G. Geuskens, Ed., Applied Science, 1975), which considers thermal, photo- and mechanical degradation.

Initially much of our understanding of polymers came from studies of their size, shape, etc., in solution. *Macromolecules in Solution* (H. Morawetz, 1975, *HP XXI*) is a survey of this field in which methods largely used to study proteins and other natural polymers are discussed along with studies on synthetic polymers. A study on a specific aspect of polymer solutions is *Light Scattering from Polymer Solutions* (M. B. Huglin, Ed., Academic Press, 1972). *Dynamics of Polymeric Liquids* (2 vols, R. B. Bird *et al.*, Wiley, 1977) deals with the rheology of polymer solutions and melts. Details of most classical methods will be found in Vol. 1 of Weissberger's *Technique of Organic Chemistry* (see Chapter 12). The classic text on the physical chemistry of macromolecules is P. J. Flory's *Principles of Polymer Chemistry* (Cornell University Press, 1953), which has stood the test of time remarkably well. The two-volume *Polymer Science: a Materials Science Handbook* (A. D. Jenkins, Ed., North-Holland, 1972) reviews properties, preparation and analysis generally rather than from a individual polymer standpoint. J. A. Brydson's *Plastics Materials* (3rd edn, Butterworths, 1975) deals with polymers individually instead. Recently translated from a German edition of 1971 is *Macromolecules* (2 vols, H. G. Elias, Wiley, 1977), which combines these two approaches.

Crystallographic Data for Various Polymers (R. L. Miller, Chemstrand Research Centre, Durham, NC, 1963) contains much useful data and is a revision of earlier tables by Miller which appeared in *Journal of Polymer Science* and elsewhere. *Introduction to Polymer Viscoelasticity* (J. J. Aklonis *et al.*, Wiley, 1972) provides a useful introduction to this topic. Although the analysis and testing of polymers is largely carried out for technological reasons, a number of fundamental studies have been made which have analytical applications. For example, *Structural Studies of Macromolecules by Spectroscopic Methods* (K. J. Iven, Ed., Wiley, 1976) discusses these techniques for elucidating the chemical and physical structure of polymers. Several hundred spectra are appended to *Identification and Analysis of Plastics* (J. Haslam, H. A. Willis and D.C.M. Squirrell, 2nd edn, Iliffe, 1972), a useful practical manual. The latest two volumes in Dekker's series *Techniques and Methods of Polymer Evaluation* are Vol. 3, *Characterization and Analysis of Polymers by Gas Chromatography* (M. P. Stevens, 1969) and Vol. 4, *Polymer Molecular Weights* (2 parts, P. E. Slade, Ed., 1975). The latter topic is also the subject of *Molecular Weight Distributions in Polymers* (L. H. Peebles, 1971, *PR XVIII*). ICI Plastics Division (Welwyn) has produced a useful 12 page guide called *Identification of Plastics for Schools* (1974), which describes simple tests such as burning. Information on the identification and estimation of polymers will also be found in standard works such as *Standard Methods of Chemical Analysis* (3 vols, F. J. Welcher, Ed., 6th edn, Van Nostrand, 1963–   ).

Mention must be made of the *Polymer Handbook* (J. Brandrup and E. H. Immergut, Eds, 2nd edn, Interscience, 1975) and its excellent bibliographies. This is a most valuable source of data and quick reference work and has sections on nomenclature, kinetics, solid state properties, solution properties, physical properties of polymers, oligomers, monomers and solvents. It has a good subject index.

The British Library Science Reference Library (London) has produced a free list of *Recent Books on Polymer Science and Technology* (1975), which does, however, exclude biopolymers, elastomers, fibres, paints and textiles.

APPLIED ASPECTS

Fibres, Films, Plastics and Rubber. A Handbook of Common Polymers (W. J. Roff and J. R. Scott, Butterworths, 1971) is useful for background information, trade names, etc. The title indicates the major applications of polymers.

Textile Chemistry (3 vols, R. H. Peters, Elsevier, 1963–76) is a useful introduction to fibre-forming polymers, while the *Handbook of Textile Fibres* (J. G. Cook, Ed., 4th edn, Merrow, Watford, 1968) and the more specialized *Handbook of Polyolefin Fibres* (J. G. Cook, Ed., Merrow, Watford, 1967) are useful quick reference sources on synthetic fibre technology and technical and economic aspects.

A German work on the production and chemistry of synthetic fibres is *Polyesterfasern: Chemie und Technologie* (H. Ludewig, Akademie Verlag, 1965), translated by B. Buck, Wiley, 1971. It is very well documented.

Unsaturated polyesters are dealt with in *Unsaturated Polyester Technology* (P. F. Bruins, Ed., Gordon and Breach, 1976) and are not used as fibres but, being reactive, are used as thermosetting plastics or can be crosslinked (cured) in the cold and so find uses in coatings and in fibre-reinforced plastics. This last topic is dealt with at length in *GRP Technology* (W. S. Penn, Maclaren, 1966), a practical guide which covers materials, equipment and processes. The equally important epoxy resins are dealt with at length in *Epoxy Resins Chemistry and Technology* (C. A. May and Y. Tanaka, Eds, Dekker, 1973). Twelve specific resins' properties are compared in the loose-leaf *Resin Properties Handbook* (Yarsley Testing Labs, Redhill, Surrey, 1976). YTL are collecting similar data on vinyl ester resins and engineering thermoplastics for publication.

In addition to their *Plastics Applications Series*, Reinhold publish *Plastics* (J. H. DuBois and F. W. John, 5th edn, 1974). *The Dictionary of Plastics* (J. A. Wordingham and P. Reboul, Newnes, 1968) is a useful source for the definition of various terms and is a companion volume to the *Dictionary of Rubber Technology* (A. S. Craig, Newnes, 1969). *The Plastics Manual* (A.E. Lever, Ed., 5th edn, Scientific Press, London, 1971) covers materials, fillers, film and sheeting, fibres, cellular plastics, technology and applications. A useful data source and international trade name directory is Saechtling and Zebrowski's *Kunststoff-Taschenbuch* (16th edn, Hanser, Munich, 1965), which lists a large number of German books on plastics. The annual subscription to the monthly *Modern Plastics* includes a copy of *Modern Plastics Encyclopedia*, a valuable source of data and (American) trade information which reviews recent progress in the applications of plastics and testing methods. *European Plastics Buyers' Guide* (IPC Business Press, annually) provides a classified guide to the UK plastics industry.

The third major grouping is the elastomers. Stern's treatise on rubbers has already been mentioned. Somewhat older but still of value is the much larger *Applied Science of Rubber* (W. J. S. Naughton, Ed., Arnold, 1961), which also covers natural and synthetic latices, reclaim, ebonite, etc., from both a theoretical and a practical standpoint.

Two more recent books on this subject are *Rubber Technology and Manufacture* (C. M. Blow, Ed., Butterworths, 1971) and *Rubber Manufacture* (M. Morton, Ed., 2nd edn, Van Nostrand Reinhold, 1973). In the important field of adhesives up-to-date treatments are provided by *Adhesion and the Formulation of Adhesives* (W. C. Wake, Applied Science, 1976) and *Adhesives. Recent Developments* (B. S. Herman, Noyes Data Corp., 1976), which is more technological, or by *Handbook of Adhesives* (I. Skeist, 2nd edn, Van Nostrand Reinhold, 1977).

A practical guide to paint formulations is *Formulation of Organic Coatings* (N. I. Gaynes, Van Nostrand, 1967). Over 200 formulations are given.

In paper technology two important works are the *Handbook of Pulp and Paper Technology* (K. W. Britt, Ed., 2nd edn, Van Nostrand Reinhold, 1970) and *Pulp and Paper Manufacture Series* (3 vols, J. N. Stephenson and J. N. Franklin, Eds, 2nd edn, McGraw-Hill, 1969–70). An equivalent work on leather is *The Chemistry and Technology of Leather* (4 vols, F. O. Flaherty *et al.*, Eds, Reinhold, 1956–65).

The polymer industry, being highly complex, generates and uses a large variety of trade literature of the type described in Chapter 15. As in the pharmaceutical industry, a vast array of products are marketed under trade names. Several reference works which contain sections on trade names have already been mentioned. Other guides are *New Trade Names in the Rubber and Plastics Industries* (RAPRA, Shawbury, 1966–) and *Trade Names: Noncellulosic Man-made Fibres, Polynosic Fibres and Textured Yarns* (ICI Fibres, Harrogate, 1976, Supplements 1976–).

Because of the importance of the mechanical properties of polymers a number of works deal with testing. J. R. Scott's *Physical Testing of Rubbers* (Maclaren, 1965) covers elastomers specifically. Some of the most extensive literature on testing methods is to be found among the various standards issued by bodies such as the British Standards Institution (BSI), the Deutscher Normenausschuss (DIN standards) and especially the American Society for Testing and Materials (ASTM). *ASTM Standards* 1978 (Parts 32–38) contain several hundred standards covering fibres, plastics and rubbers, while other parts cover paints, paper, leather, etc. Further details can be obtained from the annual *Index to ASTM Standards* (ASTM, Philadelphia). Similarly, appropriate British Standards can be located from the *British Standards Yearbook* (BSI, London). This type of literature is discussed at considerable length in E. R. Yescombe's *Plastics and Rubber: World Sources of Information* (Applied Science, 1976), which may be consulted for publication details of literature, old and new, which space does not permit mentioning here.

14

The use of the patent literature

C. Oppenheim

THE NATURE OF A PATENT

To start with, what is a patent? The simplest way of regarding a patent is as a bargain between an inventor and the community. In return for telling the community about his invention, the inventor receives a monopoly in its use for a limited period. The community benefits by the addition of more knowledge to its published stock and, of course, it can make free use of the invention after the patent has lapsed or expired. The inventor, on his side, is given the right for a number of years to stop other people making unauthorized use of his invention. In this way he has some prospect of getting a return on the time and money that went into the making of the invention. If anyone does make, use or sell the invention without permission, the patentee can sue for infringement. He will get substantial damages if he wins the case.

The concept of exchange of limited monopoly rights for information about the invention underlies the patent systems of most non-Communist countries, but the details of the bargain differ from country to country. For this reason, comments on patent law in this chapter have been restricted to only those points of direct relevance to a searcher.

APPLYING FOR A PATENT

An application for a patent can always be made by an individual inventor, but in Britain and most other countries, where the inventor is an employee, the application can be, and usually is, made by the employer. The major exception to this latter rule is the USA, where the law demands that the application shall always be made by the inventor; if he is an employee, he will then usually assign his rights in the invention to his employer.

When an application for a patent is lodged, it must be accompanied by a specification describing the invention. The specifications are usually drawn up by an expert in patent law known as a 'patent agent', acting on instructions from the inventor. The specification, an application form and a fee are then deposited at the Patent Office, or offices in the country or countries in which patent protection is desired.

Filing the application at the Patent Office establishes a certain priority in the invention described in the specification. This means that if at some later date someone else makes the same invention independently, this second inventor cannot obtain a patent if the first inventor carries on with his application properly. Obviously, from the simple meaning of the word, two 'monopolies' for the same invention cannot be granted, so time of arriving at the Patent Office becomes the criterion used to decide between two people making the same invention independently. Not only is the one who arrives too late unable to get a patent, but also he cannot even use the invention if the first one succeeds in obtaining a valid patent.

The date of arriving at the Patent Office is the important one in most countries. The major exceptions to this way of deciding priority are the USA and Canada. There, if there is any dispute on priority between two inventors, each is required to establish the date on which he conceived the invention and then to show that he was diligently developing it from the conception date until he filed a patent application.

Priority rights are also affected by an international agreement known as the International Convention for the Protection of Industrial Property. Most countries subscribe to this Convention. One of its provisions is that an inventor can, within one year of filing a patent application in any Convention country, file in any other Convention country an application which will then be treated for priority purposes as if it had been filed on the date of the first application.

One important effect of the International Convention needs to be noted: one invention can (and often does) give rise to many patents in many different countries. This has obvious implications for searching the patent literature. It is often sufficient to examine only the first of a 'family' of patents to be published (usually called the *basic*). Patents appearing subsequently (usually called *equivalents*) add to the volume of patent literature but not to the sum of knowledge. However, because of differing patent laws over the world, equivalent patents may be subtly different from one another. This problem is unlikely to be of concern to a general searcher, however.

EXAMINATION

Although patent laws vary considerably, in most countries *examination* takes place sooner or later after application. The main object of this examination is to see whether the monopoly asked for will really be for something new and not something already published before the priority date of the patent application. If it has been so published, the inventor has nothing to offer the community in return for a monopoly and he is not entitled to a patent. It is worth stressing this point that *any* prior publication of the invention, even by the inventor himself, will stop him getting a patent.

In the course of examining a patent application, most Patent Office examiners may, in theory, search all published literature for prior publication of the invention, but in practice the search is often more limited. The examiner then writes to the inventor, through the latter's patent agent, to tell him of any prior disclosures which are the same as, or closely related to, the invention under examination. The inventor (or more usually his patent agent) and the examiner then argue as to whether the former is entitled to a patent: often during the course of this argument the specification and the monopoly claimed are amended to make clear the distinction between the invention under examination and what has been previously disclosed. If the Patent Office examiner is finally convinced, the application is officially 'accepted' and the specification is published.

In some countries this will be the first time the specification has been published. In other countries (those that use the so-called *deferred examination* system) the application will have already been published before it has been examined. The patent literature has been flooded by such published unexamined applications in recent years, as more and more countries have adopted the deferred examination system. (The reasons for this trend towards the deferred examination system are too complex to go into here. The UK is one of the latest countries to commit itself to the system.) Some of these published applications will be valueless, as they represent invalid applications later to be withdrawn by the applicant or thrown out by an examiner. Because of examiners' amendments, others will bear little resemblance to subsequent patents with the same title and inventor.

As a result, the patent searcher is plagued not merely by a single invention appearing in many countries, but also by a single invention giving rise to two publications in a given country, i.e. application and subsequent patent specification.

THE PUBLISHED PATENT SPECIFICATION

Figure 14.1 shows the front page of a typical British patent specification. For most countries, two things are needed to refer to a patent specification accurately and completely: the country and the serial number. On the first point, British specifications are not particularly helpful. The wording of the specification makes clear to those well versed in patent law that the specimen illustrated is British. To the layman, however, the only indications of origin are (a) the word 'London' in the bottom of the stamp at the top left of the front page of the specification, and (b) the printers' imprint in very small letters on the last page (not illustrated): this states that the specification was printed for Her Majesty's Stationery Office and published at the Patent Office, London. Some, but not all, other countries are more helpful than this; great care is necessary in establishing the nationality of a patent specification: it should not be confused with the nationality of the applicant. The number of the specification illustrated is clearly given at the top right — 1 410 579. For reference purposes, therefore, this specification is BP 1 410 579.

There is another number in small print at the top of the specification. This number, 7935/74, is the 'Application Number' — that is, the number given to the application by the Patent Office at the time it was originally filed. The application was identified by this number until it was published by the Patent Office, when the number 1 410 579 was allocated to it. After this point in the application's life, the application number has little significance and should not be used for general literature reference purposes. In recent years most countries have adopted the standard numbering for the bibliographical information in their patent specifications recommended by ICIREPAT (Committee for International Co-operation in Information Retrieval among Examining Patent Offices). Thus, for example, the number (54) always precedes the title of the invention. This can be very useful for identifying the details of a specification.

Further discussion on the bibliographical data to be found in patent specifications can be found in the works by Finlay (1969) and Liebesny (1972).

A patent specification ends with a series of claims setting out the exact scope of the monopoly which is claimed. The general structure of sets of claims varies from country to country, but as most of the variations have only legal significance, they need not be dealt with here.

PATENT SPECIFICATION (11) **1 410 579**

1 410 579

(21) Application No. 7935/74   (22) Filed 21 Feb. 1974

(31) Convention Application No. 2604/73   (32) Filed 22 Feb. 1973 in   (19)

(33) Switzerland (CH)

(44) Complete Specification published 22 Oct. 1975

(51) INT CL² C12D 13/06; C13K 11/00

(52) Index at acceptance
C2S
C3H   Cl

(72) Inventor PETER WEBER

(54) MANUFACTURE OF GLUCOSE ISOMERASE AND CONVERSION OF GLUCOSE TO FRUCTOSE

(71) We, L. GIVAUDAN & CIE SOCIÉTÉ ANONYME, a Swiss Company, of Vernier-Genève, Switzerland, do hereby declare the invention, for which we pray that a patent may be granted to us, and the method by which it is to be performed, to be particularly described in and by the following statement:—

The present invention is concerned with an isomerisation process.

It is known that various microorganisms are capable of isomerising glucose to fructose by means of a glucose-isomerase. Examples of such microorganisms are members of the genus *Streptomyces, Bacillus, Lactobacillus, Areobacter* and *Pseudomonas.* Amongst the Streptomycetes there may be mentioned, in particular, the species *Streptomyces olivochromogenes, Streptomyces wedmorensis, Streptomyces olivaceus, Streptomyces venezuelae, Streptomyces phaechromogenes, Streptomyces albus* and *Streptomyces rubiginosus* which, after induction with xylose, are capable of producing a glucose - isomerase.

It has now been surprisingly found in accordance with the present invention that the known species *Streptomyces glaucescens,* especially the strain ETH 22794, which is known to produce antibiotics, especially hydroxystreptomycin (described in "Systematik der Streptomyceten unter besonderer Berücksichtigung der von ihnen gebildeten Antibiotika", R. Hütter, Verlag S. Karger, Basel und New York, Bibl. Mikrobiol. *6* (1967), 90—92) is also capable of synthesising a glucose - isomerase on a suitable nutrient medium. *Stereptomyces glaucescens* is distinguished as an isomerase producer vis-à-vis other known microorganisms possessing this property in that it contains a very high proportion of extra-cellular glucose-isomerase which is of advantage in the isolation.

The present invention is based on the foregoing finding and is accordingly concerned, in one of its aspects, with a process for the manufacture of a glucose-isomerase, which process comprises cultivating microorganisms of the species *Streptomyces glaucescens* in a nutrient medium and isolating the glucose-isomerase therefrom. The invention is also concerned, in another of its aspects, with a process for the manufacture of fructose from glucose, which process comprises incubating a glucose-containing solution with a culture of *Streptomyces glaucescens* or a glucose-isomerase isolated therefrom and isolating the fructose from the incubation mixture.

A strain of *Streptomyces glaucescens* which is preferably used in the present invention is *Streptomyces glaucescens* ETH 22794. (The culture collection designed herein by "ETH" is Eidgenössische Technische Hochschule Zurich).

The cultivation of the microorganisms used in accordance with the present invention can be carried out in a manner known *per se* under aerobic conditions; for example, in surface cultures or, preferably, in submerged cultures, using stirred or shaking fermenters.

A suitable nutrient medium, which can be solid or liquid, contains a source of assimilable carbon and a source of assimilable nitrogen and, advantageously, mineral salts and trace elements. Suitable sources of assimilable carbon are, for example, malt extract, glucose, maltose, saccharose and other sugars, glycerine, cornsteep liquor, amino acids, peptides, fats and fatty acids. A source of assimilable nitrogen can be microbial, vegetable, animal and inorganic nitrogen compounds such as yeast extract, peptones, bactotryptone, meat extract, amino acids, pancreatically- or acidically-hydrolysed casein, soya bean meal and sodium nitrite. Finally, in order to insure a good growth, the presence of a series of macro- and trace elements, preferably of inorganic origin, such as, for example, magnesium, cobalt, phosphorus, sulphur, is advantageous. If required or if desired, special growth factors or growth stimulants, for

[*Price 33p*]

Figure 14.1

Fortunately, from a technical information point of view, many countries' patents have the widest claim first. To obtain a short summary of what the patent specification is about, the reader could start with this claim. It should be read slowly and carefully, as it is written to define a legal monopoly and not particularly to assist a reader.

However, this claim should in no way be regarded as representing an abstract of the patent specification, and should never be copied out word for word. (The claims of BP 1 410 579 are shown in *Figure 14.2*.)

After claim 1, thereroie, the general reader should turn back to the patent's first page (*Figure 14.1*) and read the main body of the specification; here he will find a pattern which, fortunately, is followed by most patent specifications. After the standard legal recitation at the beginning, there follows a paragraph which rarely does more than repeat the heading. Then, if it is helpful for an understanding of the invention, the background of the general field in which the invention lies is sketched out, usually with great emphasis on the problems which are now solved by the invention to be described. In the particular specification shown in the figures it was not thought helpful to do this. The next feature of the general pattern is a statement of the invention, first in reasonably normal language, and then repeated in the exact legal language used in the claim. Sometimes the former is omitted, but the latter is almost always present.

Next the terms used in the claims are defined and exemplified. After all the reactants and reaction conditions have been considered in turn, there is a detailed example of how to carry out the invention; many specifications give several examples. That is usually the end of the technical information. There is sometimes a legal disclaimer inserted just before the claims in the specification.

To summarize, therefore, the standard pattern of a patent specification is usually:

General background
Precise statement of invention
Explanation of factors involved
Specific working examples of the invention
Claims

Once the reader gets used to this standard pattern, he will find it much easier and quicker to extract the information he wants from a patent specification.

The same general pattern is very common in most countries, but it is necessary to sound one or two words of warning about the claims of United States specifications.

First, there is a much lower chance than with other specifications that the first claim will be the widest in scope; in addition, with the US

In order to isolate the enzyme formed intracellularly, the mycelia were disintegrated by ultrasonic treatment while cooling with ice-water.

Example 3.

In a 1 litre Erlenmeyer flask provided with a baffle, 400 ml of a nutrient medium were inoculated with a sterile aqueous spore suspension of *Streptomyces glaucescens* ETH 22794. The nutrient medium had the following composition:

Yeast extract Difco)		
(trade mark)	10	g
$K_2HCO_4.3H_2O$	0.5	g
$MgSO_4.7H_2O$	0.25	g
Glucose	10	g

The medium was made up to 1000 ml with distilled water and adjusted to a pH value of 7.2—7.3 with sodium hydroxide. After 40 hours, the mycelia were harvested using a filter centrifuge (7500 revolutions per minute), washed with a 0.95% sterile sodium chloride solution and transferred into the induction medium which had the following composition:

Casein hydrolysate (pancreatic)	10	g
$CoCl_2.6H_2O$	0.24	g
$MgSO_4.7H_2O$	0.25	g
$K_2HPO_4.3H_2O$	0.50	g
Xylose	5.0	g

The medium was made up to 1000 ml with distilled water and adjusted to a pH value of 7.0 with sodium hydroxide. After an induction period of 52 hours, the yield of glucose-isomerase amounted to 2.2 units/ml of culture solution. The mycelia were centrifuged off, washed twice with 0.01—M phosphate buffer (pH 7.0), which was 0.001 molar with respect to $MgCl_2.6H_2O$, 0.001 molar with respect to $MgSO_4.7H_2O$ and 0.001 molar with respect to $CoCl_2.6H_2O$, and stored in this buffer solution at 4°C. The enzyme suffered no substantial loss of activity during the course of 14 days.

Example 4.

In a procedure analogous to that described in Example 3, a yield of glucose-isomerase of 3.5 units/ml of culture solution was achieved with an induction period of 120 hours. The mycelia was acetone-dried. The glucose - isomerase was isolated from the dried powder in a yield of 90% by extraction with 0.05—M phosphate buffer (pH 7.0) at 25°C.

Example 5.

In a stirred vessel maintained at 70°C, 10 litres of a solution of 1 kg of glucose, 2.4 g of $CoCl_2.6H_2O$ and 2.5 g of $MgSO_4.7H_2O$ in 0.05—M acetate buffer (pH 7.0) were treated for 4 hours with 244 g of freshly harvested and heat-stabilised mycelia (containing 21000 glucose - isomerase units) obtained from *Streptomyces glaucescens* ETH 22794. After removal of the mycelia by centrifuging, 600 g of fructose were obtained from the filtrate in the form of the calcium double salt. The separated mycelia can be used for further isomerisations of glucose to fructose. In an analogous manner and with a similarly good yield, glucose could be isomerised to fructose using 54.5 g of acetone-dried mycelia or using 91.1 g of lyophilised mycelia.

WHAT WE CLAIM IS:—

1. A process for the manufacture of fructose from glucose, which process comprises incubating a glucose-containing solution with a culture of the microbiological species *Streptomyces glaucescens* or a glucose-isomerase isolated therefrom and isolating the fructose from the incubation mixture.

2. A process according to claim 1, wherein the glucose-containing solution is incubated with a culture of *Streptomyces glaucescens* ETH 22794 or a glucose - isomerase isolated therefrom.

3. A process according to claim 1 or claim 2, wherein there is used a microorganism in which the formation of glucose-isomerase has been induced by cultivation of the microbiological species on a xylose-containing nutrient medium or on a nutrient medium containing a polysaccharide degradable to xylose by said microorganism.

4. A process for the manufacture of fructose from glucose, substantially as hereinbefore described with reference to Example 5.

5. Fructose, when manufactured by the process claimed in any one of claims 1 to 4 inclusive.

6. A process for the manufacture of a glucose-isomerase, which process comprises cultivating microorganisms of the species *Streptomyces glaucescens* in a nutrient medium and isolating the glucose-isomerase therefrom.

7. A process according to claim 6, wherein *Streptomyces glaucescens* ETH 22794 is used as the microorganism.

8. A process according to claim 6 or claim 7, wherein the nutrient medium contains, as an induction agent, xylose or a polysaccharide degradable to xylose by the microorganism.

9. A process for the manufacture of a glucose-isomerase, substantially as hereinbefore described with reference to Example 1—4.

10. A glucose-isomerase, when manufactured by the process claimed in any one of claims 6 to 9 inclusive.

Figure 14.2

method of constructing claims, it will often take the reader longer to work out which is the widest than to read the whole specification, so the value of the claims for quick reference is reduced.

The second point arises out of US patent practice. It frequently happens, in any country where the Patent Office carries out anything more than an examination of formalities, that the claims finally allowed to the applicant are narrower than those he originally asked for. In Britain, when this happens, the applicant is given the opportunity to bring the material of his specification into line with the scope of his final claims, matter that has become irrelevant being omitted. In the USA a specification cannot be amended once it has been filed, except under special circumstances. The claims, of course, may have to be narrowed, as in other countries. The net result is that in many US specifications the claims are not a good guide to all the contents of the specification.

A very useful feature of many countries' specifications is that there is printed a list of the references found in the Patent Office's literature search.

THE GRANT AND THE TERM OF A PATENT

Finding the required information in a patent specification is not the end of an enquiry, as it would be if the information were contained in general literature. Before any use can be made of the information, more aspects of patent law have to be considered.

If the enquirer merely wants to do genuine experimental work on the invention, to see, for example, whether he can make it work, he would normally be advised to go ahead, provided that he does not attempt to sell the products of his experiments. If, however, he wishes to go beyond mere experiment (he may wish to manufacture or import the item), he must first see whether the patent is still in force. This cannot be established by reading the specification unless the dates show that the patent is so old it must have expired by now. The normal way to find out whether a patent is in force is to inspect the Official Register of Patents held at the National Patent Office. Normally one can only do this on payment of a fee.

In most countries, including the UK, if a patentee wishes to keep a patent in force, he has to pay a fee each year up to a maximum number of years (somewhere in the range of 14–20 years, depending on the country). The fees are usually on a sliding scale. If the fees are not paid, the patent *lapses*. At the end of the maximum number of years the patent *expires*. In some countries, including the USA, no renewal fees are paid and the patent simply remains in force until the end of a prescribed period.

PATENTS AS SOURCES OF CHEMICAL INFORMATION

There is a widespread belief that patents are a relatively unimportant source of information. This belief is based partly on the notion that everything of importance is bound to enter the journal literature, and partly because patents are such unknown, daunting documents. They are full of legal jargon couched in a language of its own, often known as 'patentese'. However, there is no question that patent literature is of crucial importance to the chemist, for four main reasons:

(1) The chemical patent literature is of the same order of size as the chemical journal literature.

(2) It frequently represents the only source of information on a topic.

(3) Patents give more information than any other form of literature.

(4) Patents are often faster than other forms of literature.

Size of the chemical patent literature

About 500 000 patents per annum are issued world-wide. Many of these are duplicates — a British patent covering the same subject matter as a Dutch patent, for example. The proportion of chemical patents varies from country to country but very roughly one patent in four is chemical. I estimate that the number of *new* chemical patents (not equivalents) appearing in the world is of the order of 100 000 per annum. Compare this figure with an estimate, from *Chemical Abstracts* figures, of 200 000—300 000 new journal articles appearing per annum in chemistry and it can be seen that patents are a source of similar order of magnitude to journal articles.

Patents frequently represent the only source of information

A few years ago, Liebesny (1973) demonstrated that 94% of a random sample of UK patents in all subjects were not also published as journal articles either before or later. Chemical topics performed 'best' in this evaluation; 91% of the sample of 313 chemical patents could not be found as journal articles!

Assuming Liebesny's results to have some generality, one can estimate that, of the 100 000 new chemical patents per annum, 90 000 never appear as journal articles, while 10 000 also appear in the form of one or more journal articles. In other words, people who rely on the journal literature alone are missing about 90 000 potentially relevant documents per annum.

It is not certain why so few patents also appear as journal articles. They often represent relatively trivial advances in technology or science not worthy of an article. Commercial reasons may play a part – a firm may not wish to over-publicize its interest in a subject field. One reason identified by Liebesny is that a scientist in industry, if he makes a new invention, must as first priority inform his firm's patent department and co-operate in preparing a patent specification. Thereafter he can, if he wishes, write a paper, but by then he may have moved into a new subject field and be reluctant to return to his old area.

To give one real-life example of this problem: Bethlehem Steel have developed an alloy-coated sheet steel for roofing called Galvalume. This product has been the subject of over 50 patents world-wide since the early 1970s, but only one journal article had appeared on it at the time of going to press (1979) (Oppenheim and Sutherland, 1978).

They give more information

Publishers may well hesitate before entering upon the publication of a description containing dozens of line drawings or presenting several hundreds of examples illustrating a certain process, especially where the difference between one drawing and another may be very slight. In the patent world, however, the inventor is supposed to disclose everything he knows about his invention in the patent specification and he is therefore required to show the many possible embodiments of this idea in the examples and drawings of his patent. Where the non-patent literature may be showing two or three examples of a new development, the patent specification may have ten times as many. There are quite a few very voluminous patent specifications which usually relate to computers or the like. It is hard to imagine that a publisher could have been found to have printed 267 pages of specification and some 780 drawings to describe the improvements in a computing device as claimed in British Patent 749 836. A similar but more recent British Patent (1 108 800) had to be bound into four volumes. Surely these patents contain much more information than can be found elsewhere.

Earlier information

Even for those patents that also appear in the journal literature, a substantial proportion of cases have the patent as the first publication. Liebesny (1972) gives some examples.

SEARCHING THE PATENT LITERATURE

Let us, then, accept that patents are a worthy source of information. How should a chemist or information scientist go about searching the patent literature? One can either rely on abstracting and indexing services which are already used as a matter of course; primarily, of course, *Chemical Abstracts*. Or one can go to a specialist patent information service.

Many major abstracting and indexing publications in chemistry cover journal articles only. For example, *Science Citation Index* (*SCI*), *Chemical Titles, Current Contents* and *Current Abstracts in Chemistry and Index Chemicus* (*CAC&IC*). Obviously, too, computerized versions of these (*SCISEARCH* for *SCI*; *ICRS* for *CAC&IC*) will also ignore patents.

A number of abstracting services *do* cover patents, including, of course, *Chemical Abstracts*. What is their coverage like? One convenient test is to compare their coverage with Derwent's *WPI* Service, which is a specialist patent information service described later on.

Using such a method, one can split up chemical abstracting/indexing services into three categories: services which do not cover patents; services which cover patents, but inadequately; and services which give good coverage of patents. This is not to imply that a service which does not cover patents adequately is a poor service. It may have very good reasons for not covering patents, or may be so good a service in other respects that it cannot be dismissed. On the other hand, one should be aware of deficiencies. Equally, the fact that an abstracting service covers patents well is no guarantee that it covers the journal literature well.

Tables 14.1, 14.2 and *14.3* show the division of some major chemical abstracting and indexing services into the three categories (Freeman

Table 14.1 SERVICES WHICH DO NOT COVER PATENTS

Gas and Liquid Chromatography Abstracts
Electroanalytical Abstracts
Chemischer Informationsdienst
Current Abstracts of Chemistry and Index Chemicus
Chemical Titles
Metals Abstracts
Current Contents
Science Citation Index
and any computer services based on these.

Table 14.2 SERVICES WHICH COVER PATENTS, BUT INADEQUATELY

Abstracts journal	Estimated no. of patents covered in 1976	Patent no. index?	Estimated no. of patents in subject area in 1976 (from WPI)
Analytical Abstracts	Less than 30 (all British)	No	3 000
Mass Spectrometry Bulletin	30 (mainly US)	No	50
Chemical Abstracts	65 000 (in 1975)	Yes	100 000
BNF Abstracts	600 (British)	No*	1 200
Copper Abstracts	Less than 20	No	70
Ref. Zhurnal Khim.	18 000	Yes	100 000
RAPRA Abstracts	7 000	Yes	15 000

*Patents are organized in a separate section at the back of each issue.

Table 14.3 ABSTRACTS WHICH GIVE GOOD COVERAGE OF PATENTS

Abstracts journal	Estimated no. of patents covered in 1976	Patent no. index?	Estimated no. of patents in subject area in 1976 (from WPI)
Abstract Bulletin of the Institute of Paper Chemistry	6 000*	No†	5 000
Organometallic Compounds	1 800	No‡	?
World Aluminium Abstracts	2 200*	No∮	350
Lead Abstracts	200	Yes	<200
Zinc Abstracts	200	Yes	<200
Plastics Abstracts	4 100	Yes	3 800
Urethane Abstracts	500	No	?

*   Many of these patents would not be called 'chemical', as they discuss using articles made from the product.
†   Patents are organized in a separate section at the back of each issue.
‡   A new sister publication, *Catalysts in Chemistry*, does have a patent number index. Both are published by R. H. Chandler Ltd.
∮   Patents are organized in separate sections within each broad subject heading.

and Oppenheim, 1978). Particular mention should be made of *Chemical Abstracts*. Its coverage of chemical patents has improved greatly in recent years, and its patent number indexes are valuable, but foreign-language patents such as those appearing in Belgium, Netherlands, Japan and the USSR are still not fully covered (Oppenheim, 1974). Anyone who uses *Chemical Abstracts* should be aware of the deficiency.

Examination of *Tables 14.1, 14.2* and *14.3* shows that only relatively specialist abstracting journals cover patents satisfactorily. Therefore anyone wishing to search the world's patent literature comprehensively will probably need to use one or more of the specialist patent-searching services.) If the searcher is concerned with only one country's patents – e.g. the UK – services are often available from the national patent office. Fuller details are given in Liebesny (1972) or in the relevant official publications (Patent Office, 1970, 1975; Anon., 1976).) The two most important commercial abstracting organizations which specialize in patents are Derwent Publications Ltd and INPADOC.

Derwent's services can be divided into two headings: *World Patents Index (WPI)* and *Central Patents Index (CPI)*. *WPI* covers every patent specification published by 24 countries, i.e. about 13 000 documents per week. Only a few weeks after a patent is published, the printed *WPI* journal containing details of it appears. This journal can be used for searches by patentee, by International Patent Classification – i.e. by subject matter – and for identifying families of patents. Basics and equivalents are clearly distinguished. The printed indexes appear weekly, with cumulations, some on microfilm. Abstracts are not provided. *WPI* is split into sections, and the chemical section is of the most use to a chemist. *WPI* began in 1974.

CPI predates *WPI* by 11 years, and covers the world's chemical patents only. It can be regarded as the abstracts follow-up to the chemical sections of *WPI*, but in practice it is far more sophisticated than *WPI*, with a range of products to suit the needs of all organizations. At the one end, abstract bulletins with printed indexes are provided; at the other end, complete packages for current awareness – manual, punched-card and computerized retrospective searches, printed and microfilm copy of specifications, concordances for identifying families of patents, all combined with a chemical code for substructure searching. However, *CPI* is not cheap, and only the larger organizations are likely to subscribe. Both services are available for on-line searching via SDC ORBIT or via INFO-LINE.

Another service for searching patents which is only available on-line, via Lockheed DIALOG, is CLAIMS. This covers only US patents from 1950 to date, and can be searched by title words, by US classification, by patentee or by inventor.

In most countries the library attached to the Patent Office contains not only the patent specifications of the particular country, but also those received in exchange with other Patent Offices. In the UK, in addition to the Science Reference Library, many Public Libraries hold sets of specifications, and in many cases have a range of foreign specifications as well as those of the UK. Most firms of Patent Agents will provide retrospective searching services for a fee, and can also arrange the supply of specifications. In addition, a number of specialist patent-searching organizations exist. Such organizations will (for a suitable fee) carry out retrospective or current awareness searches, carry out searches for equivalents, supply copies of specifications, etc. Any foreign specification received at the British Patent Office is available for photocopying. British specifications can be purchased from the Sales Branch of the UK Patent Office.

Changes in patent law

Two conventions signed recently – the European Patent Convention and the Community Patent Convention – have implications for patent searches. The former Convention, which covers most Western European countries, requires that all these countries adopt a deferred examination system. In addition, the European Patent Office as set up by this Convention will publish its own patents. Similarly, the Community Patent Convention will lead to a new set of patent publications covering the EEC countries only. A deferred examination system will be employed. Once again, this has implications for the output of patent literature which are hard to quantify. The implications of changes in patent law on patent searching have been revealed recently (Oppenheim, 1978).

ACKNOWLEDGEMENT

I am indebted to the Controller of HM Stationery Office for permission to reproduce the portions of British Patent Specification No. 1 410 579 which are shown in *Figures 14.1* and *14.2.*

REFERENCES

Anon. (1976). *Patent Information and Documentation – an inventory of services available to the public in the European Community*, Verlag Dok., Munich
Finlay, I. F. (1969). *Guide to Foreign-language Printed Patents and Applications*, Aslib
Freeman, J. E. and Oppenheim, C. (1978). *Information Scientist,* **12,** 83
Liebesny, F., Ed. (1972). *Mainly on Patents*, Butterworths
Liebesny, F. *et al.* (1973). The Scientific and Technical Information Contained in Patent Specifications. OSTI Report 5177
Oppenheim, C. (1974). *Information Scientist,* **8,** 133
Oppenheim, C. (1978). *J. Docum.,* **34,** 217
Oppenheim, C. and Sutherland, E. A. (1978). *J. Chem. Inf. Comp. Sci.,* **18,** 126
Patent Office (1970). *Searching British Patent Literature*, HMSO
Patent Office (1975). *Patents as a Source of Technical Information*, HMSO

Further reading

Anyone wishing to learn more on patents and searching the patents literature is recommended not to go to the standard patent law works such as Terrell's *The Law of Patents* or Blanco White's *Patents for Invention*, which are meant for the expert. Instead, they should look at Liebesny's (1972) book. One should also note that in view of the major changes in UK patent law in 1977, anything published prior to that date will be unreliable for details of UK law.

WARNING

In this chapter it has been necessary to deal in outline with a number of points of patent law to provide a background to the use of patent literature. It must be emphasized that this has, of course, not been a comprehensive account of patent law. In general, someone expert in patent law should always be consulted on any problem in which the legal side of patents is involved.

15

Government publications and trade literature of interest to the chemist

R. T. Bottle and *J. S. Rennie*

TYPES OF GOVERNMENT PUBLICATION IN THE UK

British government publications are officially divided into two groups: Parliamentary and non-Parliamentary. Parliamentary publications are those necessary for, or directly the result of, the work of both Houses of Parliament and it is to these documents that most guides to government publications generally devote most space. The chemist is likely to find the second, non-Parliamentary, group more interesting; it contains publications from the various government departments and reflects the Government's ever-increasing interest and activity in a wide range of public affairs, including agriculture, health, scientific and medical research, trade, industry and social matters.

Most government publications are published through Her Majesty's Stationery Office, whose output is enormous and astonishingly varied. Some departments, such as the Ministry of Agriculture, Fisheries and Food, issue some of their publications without recourse to HMSO; the United Kingdom Atomic Energy Authority publishes its own documents and publication lists, but all priced documents may be bought through HMSO. Because of the differences in publishing policy between different departments and the fact that not every department issues its own publications list, valuable items can be missed. The interested reader may therefore find it helpful to acquaint himself with the responsibilities of the various government departments and to write to them for details of publications in addition to consulting the official HMSO lists. Suitable guides are *The Central Government of Britain (COI Reference Pamphlet No. 40)* and *Technical Services for Industry*, published by the Department of Industry. More detailed information on government activities can be found in the *Central Office of Information Reference Pamphlets*, a series of publications on home and overseas affairs.

The original division into Parliamentary and non-Parliamentary publications is no longer entirely clear, so that the chemist may find it more worth while to consider the various items in groups defined by subject or purpose.

THE WORK OF PARLIAMENT

Parliamentary proceedings are recorded in the *Journals of the House of Commons* and the *Journals of the House of Lords*, both of which record what is done rather than what is said. Less detailed records are provided by the *Votes and Proceedings of the House of Commons* and the *Minutes of Proceedings of the House of Lords*. Parliamentary debates are recorded in *Parliamentary Debates (Hansard)* in separate issues for the Commons and the Lords. The Commons *Hansard* is more generally available than the other records and undoubtedly provides the most entertaining reading.

Lords and Commons Papers are generally 'Returns' from government departments in response to a directive of the House to obtain information, 'Act' papers comprising reports or accounts required to be laid before Parliament under the provisions of certain Acts, Reports of Select Committees, and Minutes of Proceedings of Standing Committees appointed to examine Bills. Some of these may be of interest to the chemist — for example, the *Reports from the Select Committee on Nationalized Industries* and *Reports of the Select Committee on Science and Technology*.

Command Papers are so called because, in theory, they are presented to Parliament by command of the sovereign; they introduce into Parliament business which did not originate there. They include the group of policy papers known as 'White Papers', State Papers, Annual Reports, Reports of Royal Commissions, Reports of Departmental Committees, Reports of Tribunals and Commissions of Inquiry and Statistical Reports. The *Reports of the Council for Scientific Policy* and the White Paper *Framework for Government Research and Development* are likely to be particularly interesting to the chemist.

LEGISLATION

Both Bills, which are drafts of proposed Acts of Parliament, and Acts come into the category of Parliamentary publications. Bills may be Public Bills which may be of general application or Private Bills which confer special powers on companies, corporations and, occasionally, private persons. Acts of Parliament are Bills which have passed both houses and received the Royal Assent.

Statutory Instruments are, strictly speaking, non-Parliamentary publications and are 'delegated legislation' — that is, legislation not directly enacted by Parliament but consisting of regulations made by Ministers under the authority of Acts of Parliament or Orders in Council. They are of two kinds, general and local.

Much of this material is of direct interest to the chemist. For example, the *Health and Safety at Work Act 1974* consolidates much earlier legislation laying down the conditions which must obtain in factories and dealing with matters concerning the safety, health and welfare of workers. The associated statutory instruments deal with such topics as the control, keeping and use of dangerous substances and control of the emission into the atmosphere of noxious or dangerous substances. An examination of the statutory instruments listed in the *Table of Government Orders* and *Index to Government Orders* will reveal much of interest. The latter publication has an alphabetical subject index displaying such topics as 'Explosives', 'Factories, Shops and Offices', 'Food and Drugs' and 'Public Health'.

In the general area of legislation it may be of interest to note that the Department of Trade is responsible for matters arising from company legislation, supervision of the insurance industry, the insolvency service and the work of the Patent Office.

Several commercial publications give compilations of legislation with annotations on its interpretation in the courts. Some of these should be extremely useful to the industrial chemist. They include *Bell and O'Keefe's Sale of Food and Drugs* (J. A. O'Keefe, 14th edn, Butterworths, 1968), which is kept up to date with a loose-leaf service volume, and *Comprehensive Guide to Factory Law* (R. McKown, 6th edn, revised by J. Parris, George Godwin Ltd, 1976).

HEALTH AND SAFETY

The Health and Safety Executive is the operating arm of the Health and Safety Commission. It is responsible for the staffs of the health and safety inspectorates, covering factories, mines and quarries, explosives, nuclear installations, alkali works and pipelines. Together with local authority inspectors and agricultural inspectors, it is responsible for enforcing the provisions of the *Health and Safety at Work Act 1974* and the still-existing provisions of preceding legislation and regulations made under it — for example, those of the *Factories Act 1961*. This includes seeing that all possible safety precautions are taken to protect workers from occupational dangers. To this end the executive has produced a number of *Advisory Publications* such as *Mercurial Poisoning — Preventive Measures in Handling Liquid Mercury and the Removal of Contamination* and a special series on *Methods for the Detection of Toxic Substances in Air*. More general information is contained in Chapter 18.

STATISTICS

The Government Statistical Service has been much reorganized and improved in recent years and many items produced from this large concentration of statistical expertise and production of data are of considerable interest. The GSS now comprises the statistics divisions of all major departments plus two collecting agencies, the Business Statistics Office and the Office of Population, Censuses and Surveys, and the Central Statistical Office, which co-ordinates the system. The chemist concerned with industrial production and sales will find much to interest him in the *Business Monitor* series, including the report of the *Census of Production* published annually in the *Business Monitor PA* series in separate parts for each industry, and in the *Business Monitor SD* series, giving details of statistics in the distributive trades. Publications such as the *Digest of UK Energy Statistics* contain tables and charts of UK energy production and consumption of individual fuels, oils and gas reserves, fuel prices and foreign trade in fuels. Another mine of information is to be found in *Household Food Consumption and Expenditure* (National Food Survey). Statistics relating to agriculture, health and overseas trade may be of some interest.

An invaluable publication covering the interests of both departments is the weekly *Trade and Industry*, giving reliable information on everything connected with trade and commerce, including statistics, production levels, prices, tariffs, regulations and exhibitions and a regular feature on technology news.

RESEARCH AND DEVELOPMENT

Government participation in research goes back to the seventeenth century with the establishment of the Royal Observatory at Greenwich in 1675. Important developments included the setting up of the Meteorological Office in 1854, of the Department of the Government Chemist in 1842 and of the National Physical Laboratory in 1900. From this point, progress was rapid, marked particularly by the establishment in 1916 of the Department of Scientific and Industrial Research (DSIR), whose existence until 1964 symbolized a profoundly important era of government research and development. This period also saw the setting up of the Medical Research Council in 1931 and the United Kingdom Atomic Energy Authority in 1954.

The mid-1960s ushered in a period of change and reorganization, with a much wider dispersal of governmental responsibilities for research and development.

Under the Science and Technology Act 1965 central responsibility for basic civil science rests with the Secretary of State for Education and Science, who is advised by the Advisory Board for the Research Councils. Responsibility for technology rests mainly with the Secretary of State for Industry. Other government departments are responsible for research and development related to their executive responsibilities. Industrial government departments have appointed Chief Scientific Advisers to co-ordinate their research needs. Government interdisciplinary research and development is co-ordinated through meetings of Chief Scientific Advisers under the chairmanship of a Deputy Secretary (Science and Technology) in the Cabinet Office.

In 1972 the Government announced its decision in the White Paper *Framework for Government Research and Development* to extend the 'customer–contractor' principle to all its applied research and development. The 'customer–contractor' principle means that the 'customers', government departments with policy responsibility for the subject matter, determine the requirements, scope and funding available for research which they wish to have undertaken by the 'contractors', who may be one or more of the following: government research and development establishments, research associations or institutes and industrial establishments. The arrangements between them must be such as to ensure that the objectives remain attainable at reasonable costs.

The Department of Education and Science is responsible for scientific research carried out through five research councils: the Science Research Council, the Social Science Research Council, the Medical Research Council, the Agricultural Research Council and the Natural Environmental Research Council. A good deal of valuable information can be found in the Annual Reports and report series of these research councils. Examples of the Agricultural Research Council's publications are the *Reports of the Committee on Toxic Chemicals*, the *ARC Research Review* and periodical reports from the institutes. Many of the Medical Research Council's publications will be valuable to chemists, especially the *Special Reports Series* such as *The Composition of Foods* and *Memoranda* such as *Introductory Manual on the Control of Health Hazards from Radioactive Materials*.

The Department of Industry is responsible for five government laboratories or Industrial Research Establishments. The areas of research undertaken by these establishments are determined by the Research Requirements Boards which help the government to identify those areas which will most benefit by R and D support by the Department of Industry and which determine the objectives and balance of research and development programmes to support departmental policies within the broad allocation of funds to them.

Of the Industrial Research Establishments, the National Physical Laboratory has a prime responsibility for the national system of measurement and for technical aspects of standards. It also incorporates a maritime institute and carries out research into novel uses of computers. It issues a wide variety of publications, including handbooks such as *Determination of Uranium and Thorium: Handbook of Chemical Methods for their Determination in Minerals and Ores* and *Notes on Applied Science, Proceedings of Symposia* and *Mathematical Tables*. The National Engineering Laboratory, the Warren Spring Laboratory and the Computer Aided Design Centre are concerned with industrial developments in mechanical engineering, chemicals and computing techniques. The Laboratory of the Government Chemist provides analytical and advisory services to government departments. These laboratories produce a wide range of handbooks, technical papers and reports, with titles as diverse as *Investigations into the Possible Mechanisms for Eliminating Tin from Tin Cans during Refuse Incineration* and *Physical Properties of Liquids and Gases for Plant and Process Design: Symposium*.

These establishments also undertake research work on a repayment basis for British industry and overseas firms. Reports (where these are for unlimited circulation) are available from the Department of Industry's Technology Reports Centre and are abstracted in *R and D Abstracts*.

The Departments of Environment and Transport are jointly responsible for the Building Research Establishment and the Transport and Road Research Laboratory. The Building Research Establishment's series *Government Papers* covers a wide range of topics, including properties of materials, and such items as *High Temperature Studies on Individual Constituents of High-Alumina Cements* may be of interest. Similarly, the Transport and Road Research Laboratory publishes *Road Notes*, e.g. *Investigations to Assess the Potentialities of Lime for Soil Stabilization in the United Kingdom*.

The Department of Energy is responsible for the nationalized coal, gas and electricity industries and for the United Kingdom Atomic Energy Authority. The UKAEA publishes its own documents and also a *Guide to UKAEA Documents* and *UKAEA List of Publications Available to the Public*. The documents include books, periodicals, pamphlets and reports. As the resources and expertise of the authority are increasingly being used to carry out work of importance to non-nuclear industries, these publications cover a very wide range of topics, including such subjects as 'Ceramics', 'Cryogenics' and 'Desalination'. Reports which are generally available include such titles as *A Computer Model of the Photochemistry of Halogen-containing Trace Gases in the Troposphere and Stratosphere* and *The Control of X-Ray Diffracto-meters Using the Harwell 6000 Series Sub-modular Units*. *UKAEA Reading Lists* are also available.

The Procurement Executive, Ministry of Defence, engages in research for defence purposes at its own research and development establish-ments and through contracts placed with industry and the universities. It also undertakes certain research for civil purposes, including meteor-ology, civil aviation, medical research and space research. The Meteoro-logical Office publishes journals, charts and meteorological data and series including *Geophysical Memoirs* and *Reports*.

The Aeronautical Research Council, which advises both the Ministry of Defence and the Department of Industry, publishes the major series *Reports and Memoranda* and *Current Papers*.

The Ministry of Agriculture, Fisheries and Food carries out research into the causes, pathology and control of animal and plant diseases and into harmful insects, mites, fungi, birds and mammals. The Food Science division also carries out experimental work on the storage of foods, interaction of food additives and food components and heavy metal contamination. Publications of this department include periodi-cals, *Advisory Leaflets*, *Booklets* and books and, from Fisheries, *Tech-nical Reports* and *Technical Bulletins* such as *Nitrogenated Soil Organic Matter*.

INDEXES TO BRITISH GOVERNMENT PUBLICATIONS

A Daily List of Government Publications is issued. In it Parliamentary publications are listed first, those of the House of Lords preceding those of the House of Commons. Non-Parliamentary publications then follow in alphabetical order of names of government departments or sub-departments. Statutory instruments, arranged by number, end the list.

The monthly catalogue *Government Publications* is arranged in similar fashion to the *Daily List* and has, also, a list of documents sold but not published by HMSO, including those from the European Communities and the United Nations and related organizations. It also has an alphabetical index which is cumulated annually, and is issued with a loose-leaf insert containing brief descriptions of a monthly selection of books published by HMSO.

The *Annual Catalogue* is a bibliography of all government publications issued during the year and includes publications of certain British organizations for which HMSO acts as sales agent. Statutory Instruments are excluded. The index to the *Annual Catalogue* is valuable; quinquennial *Consolidated Indexes* are also issued. A supplement, *International Organisations and Overseas Agency Publications*, is published with the *Annual Catalogue*.

Statutory Instruments are listed in the monthly *List of Statutory Instruments* and an annual list is also produced. *The Table of Government Orders* issued annually lists statutory rules, orders and instruments by year and number. *The Index to Government Orders* is issued in two volumes biennially with entries grouped under subject headings. An *Index to the Statutes* is also published annually in two volumes.

For those wishing to maintain a card catalogue of official publications, a card index service is available on subscription, the cards being issued at fairly regular intervals.

Sectional Lists are catalogues of current non-Parliamentary publications with a selection of Parliamentary publications, in separate lists based mainly on the divisions of responsibility between sponsoring departments. They are free and are brought up to date regularly. *Sectional Lists numbers 1* (Agriculture and Food), *3* (Trade, Industry, Energy, Prices and Consumer Protection), *12* (Medical Research Council) and *18* (Health and Safety Executive) are likely to be of most interest to the chemist.

Keeping up with government publications can be helped by consulting the monthly news sheet *Advance Information on Government Publications*, which lists important new titles in preparation with brief descriptions and bibliographical details. Or the weekly *List of Non-Parliamentary Publications Sent for Press* can be scanned. One non-official source of information is *Keesing's Contemporary Archives*, a weekly loose-leaf service of abstracts from the press and important parliamentary papers.

Further information on British government publications is given in the following sources:

Central Office of Information (1975). *The Central Government of Britain*, COI Reference Pamphlet No. 40, HMSO
Central Office of Information (1976). *The Promotion of the Sciences in Britain*, HMSO

Cornelius, I. V. (1973). *British Government Publications: an Introductory Guide*, Stirling University Occasional Paper No. 1
Department of Industry (1975). *Technical Services for Industry*
Johansson, E. (1976). *Checklist of British Official Serial Publications*. 8th edn, The British Library
Marshallsay, D. (1972). *Official Publications: Survey of the Current Situation*, Occasional Paper No. 3, University of Southampton
Marshallsay, D. (1975). *British Government Publications*, University of Southampton
Olle, J. G. (1973). *An Introduction to British Government Publications*, 2nd edn, Association of Assistant Librarians, London
Pemberton, J. E. (1973). *British Official Publications*, 2nd revised edn, Pergamon Press

NON-BRITISH GOVERNMENT PUBLICATIONS

Many of the publications discussed by Crossley in *The Use of Biological Literature* (pages 74–83) as being of interest to the agriculturalist and biologist are also likely to be of interest to chemists. He has also given ways of monitoring and tracing the publications of the major foreign and Commonwealth countries and international organizations. Marshallsay's 1972 publication mentioned above also discusses this subject.

In the Communist countries virtually everything is technically published by the State. Excepting this special case, the US Government is the government responsible for most publications of interest to the chemist. US Atomic Energy Commission publications have already been discussed in Chapter 10 and other reports in Chapter 3. L. B. Schmeckebier and R. B. Eastin's *Government Publications and their Use* (2nd revised edn, Brookings Institution, Washington, DC, 1969) is a good basic guide to US government publications. Another, more detailed, guide is L. Andriot's *Guide to US Government Publications* (Documents Index, McLean, Virginia, 1976), now issued in four volumes, the individual volumes being revised and issued at six-monthly intervals. Popular guides include W. P. Leidy's *A Popular Guide to Government Publications* (Columbia University Press, 1968) and Ellen Jackson's *Subject Guide to United States Government Publications* (American Library Association, 1968).

The major nations maintain Scientific and Chemical Attachés at their embassies who will supply information on publications, including government ones, published in their respective countries. Their addresses can be found by consulting the *London Diplomatic List, London Telephone Directory, Kelly's London Directory, Whitaker's Almanack*, etc. A list of the various government agencies (mainly American) from which reports can be obtained appears in the editorial pages of the first issue of each volume of *Chemical Abstracts*.

The agencies of the United Nations are prolific publishers of reports, etc.; those of most interest to chemists are those of the International Atomic Energy Agency (see Chapter 10), UNESCO and the World Health Organization. These organizations produce their own catalogues. Publications of interest to chemists include the UN Statistical Office's quarterly *Commodity Trade Statistics*, the World Health Organization's *Pesticides Residues Series* and (from the International Agency for Work on Cancer) *IARC Monographs on the Evaluation of Carcinogenic Risk of Chemicals to Man*. A useful guide is *United Nations: International Bibliography, Information, Documentation* (R. R. Bowker and Unipubs Inc., 1973) and, as a monitoring service, the monthly *United Nations Documents Index*.

The publications of the European Communities are steadily growing in number and are listed monthly in *Publications of the European Communities* (in six languages, issued by the office for Official Publications of the European Communities, Luxembourg).

Certain UN and European Communities publications are listed in the monthly *Government Publications* and in the supplement to the *Annual Catalogue* published by HMSO.

TRADE CATALOGUES AND TRADE LITERATURE

These two types of literature are much used in industry, but contain much useful information even for the research worker (see Chapter 16, where their use for research and commercial intelligence is briefly discussed). Their greatest value is in saving one the time and trouble of making chemicals and equipment that are commercially available, as they are the primary source of information on these.

One reason why this type of material is not more widely used lies in the paucity of such collections, especially in academic libraries (Smith, 1966). The State Central Technical Library in Prague has, however, demonstrated the considerable increase in use which follows as more material is added to the collection and it becomes better known (Petera, 1960). There was a considerable increase in loans from 1962 onwards following the introduction of a system of automatically circulating news of accessions on a SDI basis (Derfl, 1967). The Prague collection is mainly of non-Czech material. In this country the Statistics and Market Intelligence Library has a collection of over 13 000 foreign trade catalogues which are available for loan.

Trade catalogues follow a variety of patterns. Wholesalers' catalogues contain little descriptive material; they usually consist of a list of products from several manufacturers in some simple, classified order, together with a list of prices. The manufacturers' names are not usually mentioned except perhaps where such a name might improve sales.

Catalogues are sometimes issued by trade associations and these, of course, list their members' products. A good example of such a catalogue is that of the Council of British Manufacturers of Petroleum Equipment. *British Petroleum Equipment Services* (1976 edn) is nearly 300 pages long, with numerous illustrations. The major portion of the catalogue consists of detailed notes or specifications of members' products arranged alphabetically by the manufacturers' names. The catalogue is completed by lists of the names and addresses of manufacturers and a buyers' guide (a classified list of products). No prices are given, the enquirer being expected to write for quotations after making his selection. A publication of this nature is thus much more than a catalogue, for, with its detailed specifications, its illustrations and indexes, it becomes a minor reference work.

Catalogues of individual manufacturers form the bulk of trade catalogues. They come in all shapes and sizes, unfortunately, which makes a collection of them difficult to keep in order. Many manufacturers are now keeping their catalogues up to date with a loose-leaf service. If one has built up a large collection of catalogues, it becomes necessary to adopt some method by which one can refer to these easily and quickly. Numerous methods have been devised and several have been reviewed in detail by Smith (1966), who concludes that the simplest method is the best. One simple arrangement is to use a numerical sequence numbering the catalogues 1, 2, 3, and so on, as received. If the collection includes brochures of only three or four pages, and single sheets relating to various products, then all the catalogues unable to be shelved by themselves should be kept in boxes of a suitable size and given a separate sequence of numbers. Catalogues able to stand on the shelf unaided can be given one block of numbers, say 1–1000, and the catalogues in boxes, a block of numbers 1001 onwards. An index to the collection is necessary, and this should be in three parts. The first part is an alphabetical list of the firms represented in the collection together with the number(s) of their catalogues. The second part of the index is an alphabetical list of products. This can be fairly general in nature with headings such as Electrical instruments, Optical instruments, Temperature measuring instruments, or it might be necessary to maintain an index with much more specific headings. The third part of the index should consist of a numerical list of the catalogues. This final section of the index is for checking purposes only – so that one can tell whether any of the catalogues are missing from the collection.

Some organizations prefer to subscribe to a commercial catalogues service such as Technical Indexes in this country or Sweet's Cataloging Service in the US. The *Chemical Engineering Index* (Technical Indexes,

Bracknell, Berks.) collects and classifies the catalogues from some 600 suppliers of plant equipment, services and material which are used in the chemical and process industries. Each subscriber receives microfilms of catalogues in the sections subscribed to and these are updated from time to time. A Blue Book is issued twice a year which consists of a product and trade name section, selector charts and suppliers' addresses. The charts contain breakdowns of the larger and more definable areas of instrumentation, processes, etc., and indicate the location of appropriate catalogues of those companies co-operating in the scheme and also which non-participating companies have similar products. The Blue Book is kept up to date with the monthly *CEI Bulletin. Electronic Engineering Index* and *Engineering Components Index* are similar services and a service on British Standards is also available. Similar services are available in the US from Information Handling Services of Englewood, Colorado, whose *VSMF Technical Information Retrieval Systems* cover industry codes and standards and government specifications, regulations and standards, Canadian and Japanese catalogues. TI market these services in the UK. TI are making product data and standards available through Prestel.

The dividing line between trade catalogues and trade literature is ill-defined. It is true that differences do exist. There are, for example, those catalogues which merely list commodities for sale, with no descriptive notes given, only prices. On the other hand, there are examples of trade literature which are really important monographs on specific commodities. A good example of this form of trade literature is the monograph on *Soda Ash* published by the Columbia Southern Chemical Corporation, California. Other examples of the monograph form of literature are to be found among the publications of the various research and development associations. For example, the Research Association of British Paint, Colour and Varnish Manufacturers issued *Pigment Particles: Their Character and Behaviour in Paint*, and the Copper Development Association has published works on *Beryllium, Copper* and *Copper and Copper Alloys*, as well as the quarterly *Copper*. The Aluminium Development Association issues Bulletins with titles such as *The Properties of Aluminium and its Alloys*, while the Coffee Brewing Institute in the United States publishes brochures entitled *The Chemistry of Coffee, Organic Acids in Brewed Coffee* and *A Chemical Study of Coffee Flavours*.

Many of the large firms also publish monographs which are the result of their research work. Two United States examples are *The Neoprenes*, published by Du Pont de Nemours and Co Inc., and the Carbide and Carbon Chemicals Co. publication *Synthetic Organic Chemicals*.

In this country possibly the most prolific of such publishers is Imperial Chemical Industries. Much of the material issued by ICI comes in the form of leaflets or brochures and these are collected together in loose-leaf binders provided by the publishers. A wide variety of commodities form the subject matter of such publications, e.g. *Rubber Chemicals for Cables, Sulphuric Acid: Manufacture and Uses, Pigments for Paint, Organic and Inorganic*, and so on.

Shell Chemicals is another firm issuing numerous leaflets and brochures of a similar nature, and also Laporte Industries, the latter issuing booklets with titles like *High Test Hydrogen Peroxide, Organic Peroxy Compounds* and *Sodium Perborate*. Some companies provide a catalogue of their publications, e.g. *Subject Index to Dyestuffs Division Publications* (ICI), which includes journal articles by the division's scientists.

There is also a large amount of information relating to medicinal products published by such firms as May and Baker, Boots, Ciba and ICI. Such publications normally first describe the product, then indicate when it should be used and finally cite examples of clinical usage. Many are collected together in *Data Sheet Compendium* (Assn of the British Pharmaceutical Industry, 2nd edn, 1977).

Although trade literature is voluminous, the location of suitable information on products or equipment is not particularly difficult. The obvious method is to write to the Technical Sales Manager of the larger companies manufacturing the product one is interested in and stating briefly for what purpose one wants to use it. One usually receives full details within the week, often with reprints of relevant articles, or even bibliographies, by the firm's and/or independent scientists on the product's applications. Many US companies publish leaflets which are good reviews of the uses of a specific chemical which they market, and these often contain good bibliographies of both journal and patent literature. Such sophisticated sales promotion literature is now starting to be issued by British companies. Such pamphlets on chemicals for the plastics industry are included in the list compiled by D. A. Lewis entitled *Index of Reviews in Organic Chemistry* (Chemical Society).

Becoming aware of new products and equipment is usually a matter of perusing the advertisements in the monthly 'technological glossies'. Many of these — for example, *Instrument Review* and *Chemical and Process Engineering* in this country or *Chemical Engineering* and *Modern Plastics* in the US — provide a Reader Enquiry Service by means of detachable postcards on which one just notes all the products or their code numbers which are of interest, and which one posts off to the magazine. The last few years have seen a considerable increase in the number of controlled circulation periodicals. They are entirely financed by direct or indirect advertising and are available free of charge to any 'approved reader' — i.e. someone likely to influence the purchase of the products advertised. *Laboratory News* is a British example of interest to the chemist; over 200 American examples from the chemical industries were listed and their contents analysed by Emery and Bottle (1970). Many contain at least one contributed article as well as new product information. Friend (1978) located and analysed 71 British controlled circulation journals useful to chemists. In addition to catalogues and pamphlets on specific products, a number of chemical and metal manufacturers publish house journals which often contain worthwhile articles and reviews. Examples are J. T. Baker's *Chemist-Analyst*, May and Baker's *Laboratory Bulletin, Hilger Journal*, Rohm and Haas's *Resin Review*, etc. Most of these are obtainable from the companies without charge or for a nominal subscription to cover postage. The bibliography and publication of house journals has been discussed at length and a directory of mainly British publications has been provided by Haberer (1967), while house organs and trade literature (especially those of US origins) as information sources have been reviewed in a 1961 ACS monograph reporting chemical literature symposia (Baer and Skolnik, 1961).

CHEMICAL PRICES AND SUPPLIERS

One of the first problems to overcome when one knows only the trade name of a chemical is to find its scientific name. For physiologically active substances the two most readily available sources are the *Merck Index* and '*Martindale*' (see Chapter 7). The Synonym Index of the former is particularly valuable if one fails to find the substance in the main section. Other useful drug indexes are described in pages 121–123 of *The Use of Biological Literature* (2nd edn) and in Chapter 9 of *Use of Medical Literature* (2nd edn, 1977). For other substances one should consult the Patent Office Index to Trade Marks or the Trade Marks or Trade Names Index of an appropriate Trade Directory. (It should be noted that Public Libraries are normally stronger on Trade Directories, etc., than academic libraries.) The 934 page *SOCMA Handbook* (Synthetic Organic Chemical Manufacturers Association and Chemical Abstracts Service, 1967) is another useful source. As well as trade names it gives the *Chemical Abstracts* index names and other chemical names, and names of derivatives and salts for 6300 commercially traded chemicals together with their molecular and structural formulas and Registry Numbers. The handbook contains over 20000 names, including those for 650 surfactant and resin mixtures and polymers.

Prices for laboratory chemicals are best obtained from the current catalogues of suppliers. These prices, divided by a factor of 3–5, can be used to guess the price of large quantities of commercially available chemicals if the prices cannot be found in one of the sources below.

European Chemical News prints each week a list of the current prices of some 200 chemicals in the UK, US, France, Germany, Italy and Belgium in local currency and in US dollars. Once a month the weekly *Chemical Age* publishes a list of current UK prices. There is some but not complete overlap with that of *ECN*. A very full and detailed list of US chemical prices is published quarterly by *Chemical & Engineering News*, and *Chemie-Ingenieur-Technik* contains a short list of materials prices in Germany each March and September. *The Directory of West European Chemical Suppliers* (Chemical Information Services, Ocean-side, NY) lists over 26 000 products (including synonyms) but the 1975/76 edition was found to have several gaps when checked. The fast-growing Japanese chemical industry is well surveyed in *Japan Chemical Directory* (Japan Chemical Week, annually). It contains a Buyer's Guide and Who's Who. Canadian prices are given in *Chemical Buyers Guide* (Southam Business Publns, Don Mills, Ontario, annual). Mellon (1965) lists a large number of Trade Directories, Buyer's Guides, etc., which provide information on chemical suppliers and prices in the US. Many specialist trade journals give price lists of chemicals used in their industry. German solvent and paint materials prices, for example, can be found in *Farbe und Lack*. The *Chemical Marketing Reporter* has a useful Hi-lo [sic] chemical price issue. To find the address of suppliers of bulk chemicals the *Buyers' Guide* issued each Autumn to subscribers to *Chemistry & Industry* is one of the most useful sources. The first two sections cover specific chemicals and groups of chemicals; other sections cover chemical plant and apparatus. The Advertisement Manager of *Chemistry & Industry* at one time advised readers that his staff will be glad to help them locate a source of supply of 'anything reasonably within the context of industrial chemistry'. Another source is *Where to Buy Chemicals and Chemical Plant* (Where to Buy Publishers, London, 1972). Most Trade Directories contain a list of chemicals and other materials appropriate to the particular trade and addresses of suppliers. It is, however, good practice to acquire a selection of catalogues from the suppliers of laboratory chemicals — for example, May and Baker and British Drug Houses for general chemicals, and Koch-Light, Aldrich, Emanuel, Alpha Inorganics, Eastman Kodak, Fluka, etc., for the more exotic ones. Suppliers of biochemicals, etc., are listed in Appendix 2 to Chapter 14 of *The Use of Biological Literature* (2nd edn).

SOURCES OF INFORMATION ABOUT COMPANIES

All those employed in industry are interested in their firm's position within the industry to which it belongs and in the developments by competitors within the same industry. (Such information is probably even more valuable to those contemplating joining a given firm!) It is in the financial, rather than the chemical, journals that such matters are discussed most clearly.

The first source for such information is the *Stock Exchange Year Book* (Thomas Skinner). This is an annual publication, giving the board of directors of each public company, an outline of its development, its capital structure and recent dividends. Rather more detailed and up-to-date information with 10 year statistical summaries is available on cards from Extel Services. A number of the better public libraries now subscribe to this excellent card service. The information relating to dividends can be kept up to date by reference to the daily *Financial Times*, in which are quoted the share prices, earnings and dividends of most public companies. It also deals with company reports and new developments in various industries. From time to time surveys of individual industries such as the chemical industry are issued as folio-sized supplements running to fifty or sixty pages. In these the developments of previous years are noted together with the prospects for future years. Such surveys are written from a purely financial point of view and so help to illuminate what might be inexplicable lapses in development within the chemical industry viewed solely from the viewpoint of the chemist (see also Chapter 16). The American equivalent of the *Financial Times* is, of course, the *Wall Street Journal*. A weekly distillation of financial news is provided by the *Investor's Chronicle*, where one can find company reports analysed and future prospects discussed. Selected information from Financial Times and Extel files are available through the Fintel service on Prestel (Viewdata) but is expensive relative to the above sources.

Some 11 000 major UK firms, both public and private, are listed in *Dun and Bradstreet's Guide to Key British Enterprises*. The publishers are the UK subsidiary of the well-known American firm of Dun and Bradstreet, which specializes in the collection of commercial information. The *Guide* lists business addresses, directors, sphere of trade and number of employees, and refers to parent company (if any). The second section classifies companies by trade and the fourth section shows group affiliations. The latter is the subject of the annual *Who Owns Whom* (O. W. Roskill, London). A Continental edition is published as a companion to the UK volume. The most useful source of information about companies is the *Kompass Register* (3 vols, Thomas Skinner, annually). The first volume lists 30 000 firms in the manufacturing or service industries which employ more than 50 staff. Volume II lists 35 000 products and services and has a detailed product index which enables one to compare the product range of all UK competitors. The company listing arrangement in Vol. III, geographically by county and then alphabetically by town, permits ready selection of particular types of company in a given area. Since 1967 *Kompass* has become the official register of the Confederation of British Industry. *Kompass* also covers Europe in separate sets. Further details of the above and related publications will be found in Chapter 11 of *Use of Management and Business Literature*, while Mellon (1965) lists some equivalent sources of financial data on US companies.

All UK companies, both public and private, are required to be registered at the Companies Registration Office, where there is a public search room with an index of some 8 000 000 companies. For a small payment one can look at the public file of a particular company which contains details of its articles of association and structure. *European Chemical News* provides a digest of business, technical and financial news of the world's leading chemical and allied concerns (see also Chapter 16). Also of interest are the improved weekly *Chemical Age* and *Chemische Industrie*. CAS/UKCIS covers 80 periodicals and newspapers world-wide for marketing, investment, production, legislation, etc., news on the chemical industry. *Chemical Industry Notes* (Chapter 5) and the SDI service *Chemical Business News* are produced from this data base.

A number of market research companies provide statistical and commercial data on a subscription or consultancy basis. These include the Economist Intelligence Unit and many of the larger management consultants. Chemical Data Services, an offshoot of the IPC Group, which publishes *ECN*, is now offering several continuously updated services on producers, etc. (*Chemical Plant Data*) and production, consumption, etc. (*Chemical Product Data*) on 100 basic chemicals, country audits on specific chemicals, booklets on the major chemical companies in USA, Japan and Europe, another on plastics processing machinery and finally a series on the communist countries (*East Bloc Market Data*). For example, profiles of 72 industrial chemicals and 81 major chemical manufacturers are contained in *Chemfacts: United Kingdom* (Chemical Data Services, 1977). Such services are expensive but their success is probably partly due to the highly specialized nature of this field and the difficulties and expense of small firms' information units trying a do-it-yourself approach.

REFERENCES

Baer, E. M. and Skolnik, H. (1961). *Advances in Chemistry Series, No. 30*, pp. 127–135
Derfl, A. (1967). Personal communication
Emery, B.L. and Bottle, R.T. (1970). *Gratis Controlled Circulation Journals for the Chemical and Allied Industries*, Upstate New York Chapter of the SLA, Rochester, NY

Friend, S.M.L. (1978). *M.Sc. Thesis*, The City University, London

Haberer, I. H. (1967). *Progress in Library Science*, Butterworths, pp. 1–96

Mellon, M. G. (1965). *Chemical Publications, their Nature and Use*, 4th edn, McGraw-Hill, pp. 187–190

Petera, J. (1960). *Technicka Knihovna*, (4), 49

Smith, E. B. (1966). 'Trade Literature; its value, organization and exploitation' in *The Provision and Use of Library and Documentation Services*, edited by W.L. Saunders, Pergamon Press, pp. 29–54

16

Some less conventional methods of obtaining information

B. Yates (revised by *N. J. Belkin*)

The previous chapters have, in the main, concerned themselves with published sources of information, handbooks, abstracting publications, journals, etc., all of which the seeker of information may consult. If, however, the basic facts of seeking information are analysed, then the obvious becomes apparent – that is, that the simplest, quickest and most efficient way of obtaining information on any topic is to ask someone who knows.

However true this may be, the most efficient way is not necessarily the most practical in that the information required may be located easily in a handbook or by using abstract journals, etc., so that this method would take much less time than finding someone who knew the answer. Nevertheless an approach to someone who knows has to be made sometimes and can lead to information which would not be traceable in any other way: indeed, one information centre (Yates, unpublished paper to Aslib Northern Branch, 24 October, 1967) in an assignment of the work done over a period of six months found that, in 31% of the enquiries received, the required information was not available, or was only partially available, from searching encyclopaedias, directories, books, journals, etc., and so sources other than these had to be tried. Examples which were quoted included information on particular aspects of the chemical cleaning and peeling of vegetables, where information was forthcoming from a Government department, a research association and trade associations; and information on particular aspects of metal spraying by plasma jet, where information was obtained from a government research establishment.

The points which emerge from this are (a) when one seeks information, a search of the literature is not an absolute or foolproof method of finding whether information is available or not; (b) it is possible to get information which is not available through the literature; (c) it is often necessary to approach several sources to get a complete or nearly complete answer to the problem. To be forewarned is to be forearmed; that is why the final sections of this chapter deal with methods of obtaining research intelligence.

Another most important aspect of this approach to seeking information is its cost effectiveness, the time spent for the return obtained. Anyone seeking information on the chemical properties of a certain material, having found nothing in the handbooks and textbooks, may be well advised to telephone the manufacturers and ask for the required data rather than continue with a search of the abstracts journals. *A telephone call would cost much less than the cost of the time the enquirer would take in his continued search.* Similarly, after, say, half-an-hour's literature search which has turned up no tangible lead an enquirer may find that a telephone call to one of the information sources covering the subject of interest would be an effective way of obtaining advice on the best sources to search or maybe even obtaining the required information. In any case it is certainly a better method of working than ploughing blindly through abstracting journals.

Much valuable information of a trade, statistical and scientific nature is accumulated in offices and laboratories, some in written form in the way of reports, pamphlets and trade literature, some in the way of 'know-how', expertise and experience. The bulk of this heterogeneous mass of information is neither catalogued nor abstracted, but yet it may be available to anyone who calls upon it, the problem being to find where it is located, since the holders and generators of such information are many and varied. It is fortunate that some useful guides exist to enable these sources to be traced, whether the source be industry, government department or laboratory, research association or independent laboratory, trade association or learned society, university or technical college, at home or abroad.

Industry is probably the prime generator of information, and for this reason it is usually the most profitable avenue to approach for assistance. One essential point to be borne in mind is that it is important to state at the outset of any such approach the organization asking for assistance. Provided that one is prepared to do this, then there need be no hesitation on the grounds of commercial secrecy because the firm being approached, knowing who is asking for information, can consider whether to release it or not, and also because information is not being demanded but is being requested. What one in fact is saying is: 'If you could help with this problem we would be grateful . . . if you cannot, perhaps you can suggest other sources who might be able to help.' This dual approach should always be borne in mind and often leads to other useful contacts.

LOCATING SUITABLE CONTACTS

Industrial Research in Britain (8th edn, Hodgson, 1976) contains a compilation of research and development activities of companies in Great Britain, though it should be noted that by no means all companies undertaking research and development are included. Some firms failed to reply to the publisher's request for information and some were excluded at their own request, for security or other reasons. In spite of the title, by far the greater part of this directory is devoted to university and technical college research, also including selective directories of government departments, other official organizations and public corporations engaged in research, research organizations, trade and development associations, research contractors and consultants and learned societies and institutes. All of these organizations, as well as the scientific-based industrial concerns from which the title is derived, could be useful in answering particular enquiries. The directory is especially valuable in that it gives names of research directors, heads of university departments and secretaries of institutes, etc., as applicable, making contact with the appropriate person easy. The descriptions of research activities can be good, but are dependent upon the individual replies. Though there is no subject index, there is a name index, which can be useful if the name of the organization is informative. Otherwise, access may require going to a trade directory for the appropriate name, and then to this directory. *Industrial Research Laboratories of the United States* (15th edn, Bowker, 1977) provides an alphabetical listing of over 5000 laboratories in the USA engaging in industrial research. Access to this directory is enhanced by an excellent subject index, plus personnel and geographical indexes. The entries contain name, address and telephone number of the organization, its principal officers and professional staff (with qualifications), and the principal areas of research. The National Referral Centre for Science and Technology in Washington, DC, also acts as a supplier of names of possible information contacts in the USA and in 1967 it published a useful, but now dated, *Directory of Information Resources in the United States*.

When one seeks information on the uses, properties, analytical methods, etc., associated with a particular material, should a suitable contact not appear using the previous sources, it then becomes necessary to use trade directories such as the annual *Chemical Industry Directory and Who's Who* (Benn, latest 1978), *Chemical Week Buyers'*

Guide (McGraw-Hill), etc., or one of the Kompass directories (Kompass Register) which are available for many European countries, or to contact the Chemical Industries Association, 3 Albert Embankment, London SE1, requesting the names of manufacturers. (See also Chapter 15.)

It is often valuable to have a good general overview of how science and technology is organized in the country before trying to locate specific sources to help with a problem. The *Guide to Science and Technology in the U.K.* (5th edn, Hodgson, 1971), edited by S. E. Macreavy, is a good, if somewhat dated, work of this sort. Although names and addresses of organizations are not included, the discursive treatment of each is generally quite detailed, giving much more information about the type of research being carried out than does *Industrial Research in Britain*. The former might therefore be a good check on information obtained in the latter. The *Aslib Directory, Vol. I: Information Sources in Science, Technology and Commerce* (4th edn, Aslib, 1977) can be valuable for contact addresses. It gives addresses and telephone numbers of member organizations and the name or title of the person within that organization to whom information enquiries should be directed. It is perhaps not as useful as the 3rd edition (Aslib, 1968), which had more complete entries and, perhaps more important, was organized alphabetically by city rather than name of organization, as is this edition. Both have subject indexes, though that to the 3rd edition is rather easier to use.

The range of government departments and their associated laboratories is far wider than is generally recognized, as are the subjects which they cover (see Chapter 15). Because of continuing changes in the structure of the UK governmental administration, it is often difficult to find just where relevant research is being done, and especially under whose auspices. At the time of writing, the government departments most concerned with relevant scientific research were: the Department of Energy, being responsible for, among other agencies, the United Kingdom Atomic Energy Authority; the Department of Education and Science, which oversees the operations of the universities and the research councils; the Ministry of Agriculture, Food and Fisheries, which maintains a number of research establishments; and, of special interest, the Department of Industry, which includes the National Physical Laboratory, the National Engineering Laboratory, the Warren Spring Laboratory, the Computer-Aided Design Centre and the Laboratory of the Government Chemist among its research establishments. In general, it will be valuable to compile the annual reports, etc., of laboratories or departments whose work might be of special interest, since there is often no other means to discover just what they are doing.

Information about research activities, as well as advice, can often be obtained from the headquarters of the departments, as well as from the laboratories. There are a number of sources which give addresses and names of contacts for government departments and their associated research laboratories.

Industrial Research in Britain provides bare data, or an approach can be made through an appropriate library, details of which can be found in the *Guide to Government Department and Other Libraries and Information Bureaux* (22nd edn, British Library, 1976). But the best source for this type of information is *Technical Services for Industry* (Department of Industry, 1975), which is a guide to technical information and other services available from government departments. It gives names and addresses of all the appropriate departments, together with suitable contact personnel and descriptions of the responsibilities and activities of the organization. This publication is especially valuable as a source of information on research organizations (see below).

The Department of Industry (DOI), being concerned with providing advice and information to industry, administers a number of programmes which are helpful in answering research enquiries or in directing them to appropriate sources. The regional offices of DOI will provide such advice and assistance, and the Small Firms Information Centres, located at ten of the regional offices in England, Scotland and Wales, will give a wide range of free information services to small firms. The Department of Commerce administers a similar program in Northern Ireland.

Interlab is another good source of direct information, and is established under the direction of DOI. It is a voluntary organization of some 2400 research and development departments of all types, set up to facilitate interchange of information by personal contact, and to make special expertise known and available. Membership (free) is open to any interested, appropriate organization. Interlab is organized on ten regional groups, each with a separate directory listing expertise and special scientific equipment. The directories are issued free to members. Interlab aims to be a service for consulting experts on specific problems, for obtaining advice on equipment problems, and for providing reference to consultants or contract research organizations.

Research associations have existed in the UK since 1918, when the British Scientific Instrument Research Association was established, and now there are over 40 such associations, with interests ranging from baking to hosiery and from brushes to paint. These associations are financed by contributions from industry, together with government support from the DOI, and each association conducts research and development work of benefit to that industry. The research and development work and background skills are paid for by members' contributions, and it is only natural that members should have prior call on the information available, but nevertheless some research associations are willing to provide information to non-members.

Some research associations also have special services which are available to anyone: for instance, the British Iron and Steel Research Association has a Steel User Service dealing with the selection and treatment of steels, and its Corrosion Advice Bureau provides advice on the corrosion of ferrous metals in air, water and the soil. The British Scientific Instrument Research Association has the SIRAID service available to anyone, and this service provides practical advice on measurement and control problems and information on where to buy British instruments and control equipment. The best source for details of the research associations is *Technical Services for Industry*, though *Industrial Research in Britain* is also useful. Similar organizations in Europe are listed in other guides considered later in this chapter.

Various other organizations have set up research and development laboratories without government support; examples are the Natural Rubber Producers Research Association and the Cement and Concrete Association. The basic aim of these organizations, set up by particular industries, is to do work which will assist in the promotion of a particular material, and this being so, information is freely given on request. The names and addresses and other relevant information on this type of research organization are included in *Industrial Research in Britain*.

Universities, polytechnics and colleges of technology should not be overlooked as sources of information, as a consideration of their number and associated expertise will show. It needs to be stressed, however, that it is not the prime function of the staff of a college or university to act as information sources and that it is often difficult to speak to the person required, certainly by telephone, because of their lecturing commitments. Nevertheless any reasonable approach, in this instance usually by letter, is likely to be answered quite readily. Some useful guides are discussed under the heading 'Research Intelligence' but *Industrial Research in Britain* gives good general information and is a useful first recourse.

Trade associations and learned societies can be useful sources of help but possibly quite naturally the amount of assistance given varies considerably. Some trade associations have technical advisory facilities, some have not, but the point to be remembered, as with learned societies, is that there are many of them and they may have the information required or be able to suggest where it can be obtained. The names and addresses of trade associations, such as the Association of British Manufacturers of Agricultural Chemicals, the National Industrial Salvage and Recovery Association and the Chemical Industries Association, can be found in *Trade Associations and Professional Bodies of the United Kingdom* (5th edn, Pergamon Press, 1971), compiled by Patricia

Millard. The arrangement of this compilation in broad subject groupings as well as alphabetically is most useful, as are the subject and geographical indexes. Alternatively, *Directory of British Associations and Associations in Ireland* (5th edn, CDB Research, 1977) can be used, or for the USA, *Encyclopaedia of Associations: Vol. 1 National Organisations of the U.S.* (12th edn, Gale, 1978).

Scientific and Learned Societies of Great Britain (61st edn, Allen and Unwin, 1964) provides a very broad subject grouping of societies and details can also be located in *Industrial Research in Britain* but no complete up-to-date compilation is available. This also applies to the USA and Canada, the most detailed available listing being *Scientific and Technical Societies of the United States and Canada* (7th edn, National Academy of Sciences, 1961). However, the annual *National Trade and Professional Organizations of the U.S. and Canada and Labor Unions* (14th edn, Columbia Books, 1979) does list most scientific, professional and learned organizations in the USA and Canada, with managing directors, a few details on the organization, and an historical note. There is a key-word index, which can be helpful.

A few methods worthy of trial when seeking information remain. One is to write, asking for help, to the editor of a periodical which is likely to cover the point of interest; and the second method is to write to the author of that article, found from a literature search, which comes nearest to answering the particular question in mind, asking whether he has extended his work to cover this question or whether he knows anyone who has.

OVERSEAS SOURCES

Some of the sources which may be used to locate information in foreign countries have been mentioned, but there are also other guides, such as the *Guide to European Sources of Technical Information* (4th edn, Hodgson, 1976), compiled by C. H. Williams, which is arranged in separate sections for each of the main technical industries (e.g. Chemical Industry), subarranged by country and then by original language title. This guide is selective and covers 'only the main sources of information in each particular field . . . which might reasonably be expected to answer enquiries from abroad . . .'. Eastern Europe is included, but in these cases the organizations are primarily trading companies which will only give information about their own products. There is a permuted name index (in English) and an original-language name index, but no other subject or geographical approach than the organization of the guide. *European Research Index* (4th edn, Hodgson, 1977) is very useful, containing a comprehensive list of research establishments in many countries, and also covering the research facilities of many European industrial firms. Arrangement is by country, with indexes both in the original language and in English. These indexes are slightly permuted, so there is some subject access. This is currently a better source than many of the country or area directories such as the *Scandinavian Research Guide* (2nd edn, Scandinavian Council for Applied Research, Copenhagen, 1965). The annual *World of Learning* (29th edn for 1978–79, Europa, 1978) is another useful source, being comprehensive in scope and covering universities, colleges, learned societies, major libraries and research institutions throughout the world. It is naturally a rather time-consuming job using the latter two sources if one cannot first specify a country likely to contain a useful source of help. Information from abroad can be obtained from the scientific attachés stationed in particular countries. The function of a scientific attaché includes reporting on scientific advances in the country in which he is stationed and the UK has attachés in such countries as Russia, Germany, France, Japan and the USA. The initial approach for information from these sources should be through the Department of Education and Science, in London, because a report may already have been sent on the topic of interest and also so that the work of the attaché can best be directed and coordinated. There is a small section on British scientific attachés overseas in *Industrial Research in Britain*, and a section on foreign embassies and scientific attachés in London.

Other countries have officers carrying out a similar function in their embassies who may be able to advise an enquirer requiring information about scientific work in their own country, though this method is by no means a very successful one, but for trade mark information and information on the manufacturers of that country's products the embassy commercial departments are usually quite good.

It will be seen that there are many sources of help other than the literature, but when one uses these sources it is necessary to keep several points in mind. Defining the problem specifically is a prime necessity when making approaches for information. The question should be expressed completely. If information is wanted on the effects of catalysts on lubricating oils, what catalysts, what temperature, what lubricating oils and what other conditions apply? Be prepared to say why the information is required, sources of help already tried, the urgency of the enquiry and, above all, who is requesting the information.

It has been found a practical proposition in most cases to make the initial approach by telephone, followed if necessary by a letter. This method at the very least gives one an immediate idea whether information is likely to be forthcoming.

Maybe even after one has searched the literature and tried the sources outlined above, information still is not available. Then the alternatives are to do without it, to do the necessary work yourself or to ask a consultant to do it for you. Consultants can be found for every branch of chemistry from sections in some trade directories and from the Royal Institute of Chemistry's *Directory of Independent Consultants in Chemistry and Related Subjects*. Entries in this directory are restricted to practices controlled by Fellows or Associates of the Institute, and an alphabetical list of names notes their special fields. A subject guide is also provided, as a help in finding practices specializing in any particular field. The *Register of Members* of the Union Internationale des Laboratoires Independents (U.I.L.I., latest 1978) is a good source for private laboratories and consultants, listing all of the members by country, their activities being described in the vernacular plus German and English. There is a subject index in English (with French and German equivalents), a product index and a 'field of investigation' index, all of which are quite useful. For the UK there is the *Register of Consulting Scientists* (4th edn, Fulmer Research Institute, 1978), which is useful but, unfortunately, neither guarantees that those listed follow any professional code of practice (as do the previous two directories) nor takes responsibility for the information in the entries. As well as sections on independent consultants and contract research organizations, there are sections on organizations which give free advice or help, on government-subsidized laboratories and research associations, and on university departments offering consulting and contract research facilities. The subject and name indexes are useful. Many university, polytechnic and technical college lecturers will do work and give advice on a consultancy basis, for which a consultancy fee, similar to that of a consultant in professional practice, would be required. Details of academic departments which will undertake work of this type can be found in both *Industrial Research in Britain*, which gives names of liaison staff and the type of facilities available, and the *Register of Consulting Scientists*.

RESEARCH INTELLIGENCE

Gathering information on research activities can be a highly specialized activity and it is one with which most bench chemists do not concern themselves. The new graduate should, however, be aware of what can be found out about the multi-faceted background to his work, and so a few pointers to this information are given in this section, though it is well recognized that these are few among many and that this complex subject merits a monograph to itself.

Academic institutions

Research work being undertaken at academic institutions can be traced without too much difficulty, since such institutions naturally tend to be less reticent about their work than industrial firms. One can find out who is doing work, as yet unpublished, in one's field of interest and in which establishment, and this could lead to an informative exchange of visits or correspondence and possibly save duplication of work.

A very good guide, with an author and subject index, is *Scientific Research in British Universities and Colleges* (HMSO), annually. It now appears in three volumes — Vol. 1, Physical sciences; Vol. 2, Biological sciences; Vol. 3, Social sciences — and it is arranged by university or college and subdivided by department, showing the work being undertaken and the person responsible.

Lists of theses and annual reports of the universities also provide a good guide, as do the lists of recipients of research grants from the Science Research Council. These lists, which are put out by the organization providing the grant, name the recipient, the scope of his work and where it is to be undertaken. Aslib produce an *Index of Theses Accepted for Higher Degrees in the Universities of Great Britain and Ireland*. This and other guides are discussed in Chapter 3.

Directory of Graduate Research (ACS, latest edn, 1977), a biennial publication, gives details of work being done in United States and Canadian universities. This lists Ph.D. theses, staff members, and their research interests and publications for the past two years, of departments of chemistry, biochemistry and chemical engineering. The only subject access is through the arrangement of the entries under the four main headings of chemistry, chemical engineering, biochemistry and pharmaceutical/medicinal chemistry. Covering the programmes of associated institutes is *Research Centers Directory* (6th edn, Gale, 1979), edited by A. M. Palmer, a guide to university-related and other non-profit research organizations. It is arranged by broad subjects, with name, institution and subject indexes. *New Research Centres* is issued by Gale Research as a supplement to *Research Centres Directory*.

The research institutes of various European countries which may be of interest can be found in some of the publications noted elsewhere, e.g. *European Research Index, The World of Learning, Guide to European Sources of Technical Information*, and in some cases in a series of reports prepared by the Organization for Economic Co-operation and Development (OECD) *Country Reports on the Organization of Scientific Research. Chemistry in the Soviet Union* (J. Turkevich, Van Nostrand, 1965) contains a useful survey of the major research institutes, biographical sketches of the older living chemists and a historical survey from 1850. The *Guide to Science and Technology in the U.S.A.*

(Hodgson, 1973), edited by D. Skevington, and the *Guide to Science and Technology in the USSR* (Hodgson, 1971), edited by S. White, are good general descriptions of how science and technology are done in those countries, with some useful addresses. Both these works follow the pattern of the same publisher's *Guide to Science and Technology in the U.K.*

Government research

The governments of the advanced countries maintain well-equipped research laboratories which undertake research work for the common good, and the work of these laboratories is usually well publicized and freely available. Most of the laboratories issue annual reports and have open days. An indication of the material available from government and international sources and its scope is given in Chapter 15.

Perhaps the most useful single source to information about scientific activities in various countries is the 24 volume series, *Guide to World Science* (Hodgson, continuing). Each volume contains extensive essays on the various aspects of the organization, financing and practice of science and technology in a geographical area (e.g. Scandinavia, Vol. 7) or country (e.g. USA, Vols 22 and 23), and a directory of the major research establishments. The series concentrates almost entirely upon governmental research and governmental sponsorship of research, but includes some non-profit research organizations as well. The sections on scientific and technical information sources are usually valuable.

Industrial research

Information about industrial research programmes is not readily obtainable and can often only be got by inference. *Industrial Research in Britain, European Research Index* and *Industrial Research Laboratories of the United States* contain details of the research activities of some firms together with the name of the research director and in many cases the number of qualified staff employed. Useful information on government and academic research is also contained in these publications.

In the main one must rely on what information can be gleaned from newspapers and periodicals. Articles such as the ones which have appeared in *Chemistry & Industry* on industrial research establishments are invaluable, and as well as articles of this nature informative news items appear in this type of publication, e.g. *Chemical Age, Chemical Processing, European Chemical News*, etc. The *Financial Times* should not be overlooked, as this is indeed a major source for anyone seeking research intelligence, both for its Technical Page feature, which runs every day except Saturday, and for the reports on all aspects of the chemical and allied industries throughout the world which it carries. Other newspapers such as *The Times, Guardian* and *Daily Telegraph* also print occasional articles of interest to the chemist. (A bi-monthly index for *The Times* is published.)

Another source of information on the work which public companies are doing is through the statement, usually made annually, by the chairman to the shareholders of the company. In the statement it is the custom to review the events of the past year and to assess future prospects. The financial newspapers usually carry details of the chairman's speech or, alternatively, a copy can usually be obtained from the company secretary.

House journals should not be overlooked (see Chapter 15); the better type communicate information on progress of research and results, and on new products, and sometimes list freely available company technical reports. The searcher should bear in mind that house journals vary considerably in their information value and should be able to distinguish between those compiled for internal use, those compiled for prestige purposes and those whose function is to act as a method of useful communication. As well as the more conventional news in this type of publication, internal news may prove valuable, e.g. appointments and promotion, and, indeed, the advertisements which a firm puts out for staff can be useful to a trained seeker of research information.

Trade literature and trade catalogues can sometimes be the first indication of some new development, providing information on properties and uses of some new chemical or of some new uses for existing materials. The information content of this class of material varies immensely, some publications being detailed technical notes, while others are only lists of materials. This type of publication has been considered in detail in Chapter 15, where it is pointed out that it is fairly easy to get copies of trade catalogues and other trade literature by writing to the company concerned but that it is not so easy to find a comprehensive set of up-to-date catalogues which is available to the general public.

Most of the larger companies produce attractive brochures covering the career prospects for prospective entrants into the company. These are useful sources of information both of a background nature and as teaching material. The Cornmarket Press distributes annually to graduates its *Directory of Opportunities for Graduates*, which summarizes such information on our major companies, research associates, civil service, etc. Companion volumes for Canada and France are also published.

Research Index (Business Surveys, Wallington) provides a guide to British information of a commercial nature and is arranged both by company and by subject. Published fortnightly, it covers both periodicals and newspapers, is available in most large public libraries, and has proved itself to be a most valuable working tool.

Source material on the American chemical scene includes *Chemical Week* (which contains information on company activities, business topics and market information as well as technological information), *Industrial & Engineering Chemistry, Chemical & Engineering News*, etc., and the less readily available *Chemical Market Abstracts* and *SEARCH. Public Affairs Information Service* (New York) is useful in being published weekly and cumulated annually. It has entries on the chemical industry in various countries and also on particular companies and particular sections of the industry. The source material used is predominantly American, selected articles from more than 1000 periodicals, book pamphlets and government documents being included. *Business Periodicals Index* (Wilson) indexes predominantly American periodicals and covers the administrative, marketing and financial side of the chemical industry. It is good for company information, as it indexes such periodicals as *American Druggist, Chemical Week, Drug & Cosmetic Industry* and *Oil, Paint & Drug Reporter*.

Anyone interested in market research intelligence is certainly interested in the names of suppliers of chemicals, prices, production figures and details of the history and affiliations of the supplying companies. Publications of interest such as *The Chemical Industry* (OECD, Paris), trade directories, etc., have been covered in detail in Chapter 15, and reference should be made to them there. A thorough survey of the topic is given in *Chemical Market Research* (N. H. Giragosian, Ed., Reinhold, 1967), which was sponsored by the US Chemical Market Research Association and covers methodology, economics and management aspects.

Banks and investment houses occasionally review particular sections of industry, e.g. *The Chemical Industry in some European Countries*, prepared in 1967 by the Economic Research Group of the Amsterdam-Rotterdam Bank, Deutsche Bank, Midland Bank and Société Générale de Banque (Belgium). Unfortunately, there is no sure way of keeping informed of or tracing such publications. Neither is there any sure way of keeping track of the numerous other publications from many sources which would help the seeker for research intelligence and company information to increase his background information. One can follow the regular series of production and import and export figures produced by the Board of Trade, Customs and Excise, United Nations, etc., and the regular surveys made by the OECD Chemical Products Committee, but the once-and-for-all type of publication, particularly by international organizations, trade associations and various committees, often does not appear in any of the abstracting journals.

Consider, for example, the series on the East European Chemical Industry produced by Joseph Crosfield and Sons Ltd, Warrington. Among the pamphlets issued in the series were *Caustic Soda and Chlorine in the Soviet Union, Costs and Product Distribution in the Hungarian Chemical Industry* and *Poland's Trade in Chemicals 1958*. Another example might be the United Nations Economic and Social Council Economic Commission for Asia and the Far East's publication, *The Development of Basic Chemical and Allied Industries in Asia and the Far East*, the proceedings of a seminar held in Bangkok, 3–13 October, 1962.

Useful comparisons regarding staffing of projects and productivity in the British chemical industry as against the American industry could be made from the National Economic Development Office's *Manpower in the Chemical Industry: a Comparison of British and American Practice* (HMSO, 1967), though the details are now out of date. Again, such publications are not easily found via abstracting journals.

Current biographical details

Biographical details of a particular chemist are occasionally needed, or sometimes just his address is required. On such occasions the lists of members published by learned societies are useful. The *Royal Institute of Chemistry Register* (latest edn, 1976), the *Royal Society Yearbook* and the membership lists issued from time to time by the Chemical Society and by the American Chemical Society are cases in point. International organizations sometimes sponsor directories, e.g. *International Union of Crystallographers Membership Directory* (4th edn, 1971), while all academic staff in Commonwealth universities are included in the *Commonwealth Universities Yearbook* (ACU, London), a four-volume annual publication. *The World of Learning* (Europa, London) and *Minerva Jahrbuch der gelehrten Welt* (W. Schuder, Ed., De Gruyter, Berlin, annually) list only professorial staff.

If more than academic qualifications and addresses is required, then publications of the *Who's Who* type are needed. For example, the *Who's Who of British Scientists, 1971/72* (Longmans, 1971) is an adequate, though out-of-date, source for the UK. *American Men and Women of Science* (7 vols, 13th edn, Bowker, 1976) is the most complete directory of this type available, arranged by broad subject headings. It is limited to the natural sciences, with engineers to be listed in a separate directory. *Who's Who in Atoms* (6th edn, Hodgson, 1977) and the three-volume *Who's Who in Science in Europe* (3rd edn, Hodgson, 1978), which has nearly 50 000 entries, give, in addition to the name and address of the scientist, information about qualifications, recent appointments, membership of societies and scientific interests. A special number of *Hamdard Medical Digest* (1967) is devoted to 'Men of Science in Pakistan, 1967'. Other directories of this type are noted in *Chemistry & Industry*, etc., from time to time.

It can be seen that research intelligence is not very easy to obtain but that there is usually much more available than appears at first sight.

17

History and biography of chemistry*

J. L. Thornton and *R. I. J. Tully*

The origins of chemistry extend far back into antiquity, and despite recent studies of the development of the subject in ancient and mediae-val India, in ancient China, among the early Greeks and in Arabic literature, much remains to be accomplished within this fascinating field. The beginnings of modern chemistry are difficult to determine, for many contributing extensively to the subject persisted in their belief in alchemy. There is no definitive period between the latter and the birth of scientific chemistry, and the overlap stretches over a considerable period.

Theophrastus Bombast von Hohenheim, generally known as Paracelsus (1493–1541), was a pioneer in experimental chemistry, and has been called the founder of iatrochemistry, or chemistry applied to medicine. Born at Einsiedeln, near Zurich, Paracelsus regarded himself as a German, and during his colourful life he travelled throughout Europe and engaged in controversy over a variety of subjects. He was qualified in medicine, and studied alchemy and astrology, and his numerous writings contain a mixture of mysticism and the results of experimental research. His chemistry is mostly found in the nine

*Within the limits of this chapter only the most significant contributions in successive chronological periods can be mentioned. Fuller information on indi-viduals mentioned, and on editions of their writings, is available in the monu-mental *A History of Chemistry*, by J. R. Partington (Vol. 1, Part 1, 1970; Vol. 2, 1961; Vol. 3, 1962; Vol. 4, 1964). Selected lists of general histories of science, of histories of chemistry, and of biographies and bibliographies of chemists, arranged alphabetically by subject, are appended to this chapter.

genuine of the ten books forming his *Archidoxis*, written about 1526 and first published in Cracow in 1569. Collected editions of his writings were edited by John Huser (10 vols, Basle, 1589–90) and by Karl Sudhoff (14 vols, 1922–33). Paracelsus was the outstanding personality of the sixteenth century contributors to chemistry, and his writings greatly influenced the work of his successors. Johann Baptista Van Helmont (1579–1644), for example, was a firm believer in alchemy and based his ideas on the works of Paracelsus. A native of Brussels, Van Helmont also qualified in medicine, but his experimental work was concerned with ferments, gases, metabolism and the physiology of digestion. He conducted chemical experiments on urinary calculi, and Boyle was influenced by his work. Van Helmont's manuscripts were published after his death as *Ortus medicinae, id est initia physicae inaudida* (Amsterdam, 1648 and 1652) and his complete writings were issued as *Opera omnia* (Frankfurt, 1682 and 1707).

Little is known about Jean Beguin (*c.*1550–*c.*1620), a native of Lorraine, who taught chemistry and pharmacy in Paris. He produced a manual of chemistry for his students, printed as *Tyrocinium chymicum* in 1610, and first commercially published in Paris in 1612, after a pirated version issued from Cologne in 1611 caused Beguin to publish this authentic revised edition. A French edition, *Les élémens de chymie* (Paris, 1615), was followed by translations into German, and into English by Richard Russell in 1669.

William Davidson (born *c.* 1593 in Aberdeenshire) became the first occupant of a chair of chemistry in France, that at the Jardin du Roi in 1648. Davidson wrote an early textbook on chemistry, *Philosophia pyrotechnica* (4 parts, Paris, 1633–35, and 1640) and *Les élémens de la philosophie de l'art du feu ou chemie* (Paris, 1651, and 1657).

The Hon. Robert Boyle (1627–91) contributed extensively to physics, and was the author of a large number of books and papers on a variety of subjects. Of particular interest from our viewpoint are *The sceptical chymist: or chymico-physical doubts & paradoxes* (London, 1661); *Tracts . . . of a discovery of the admirable rarefaction of the air* (1670); *Tracts . . . containing new experiments, touching the relation between flame and air* (Oxford, 1672); *Of a degradation of gold made by an anti-elixir: a strange chymical narrative* (1678); *The general history of the air* (1692). His *Works* were published in five volumes, 1744, six volumes, 1772, and a Latin edition in three volumes (Venice, 1696–97).

The son of another alchemist, Johann Kunckel (1630 or 1638–1703), of Hutten, near Rendsburg, conducted experiments on metals and glass, and was the author of several books in German but with Latin titles. Probably the most important of his writings was the posthumously published *Collegium physico-chymicum experimentale, oder*

Laboratorium chymicum (Hamburg and Leipzig, 1716). Andreas Libavius (1540?–1616) was born at Halle and became director of the Gymnasium at Coburg. His many publications at Frankfurt include a collection of letters on chemical subjects, *Rerum chymicarum epistolica forma ad philosophos et medicos quosdam in Germania excellentes descriptarum* (1595–99); the first systematic textbook of chemistry and his most important work, *Alchemia* (1597); and *Alchymistische practic* (1603).

A critic of alchemy but himself a controversial author of far-fetched theories, Athanasius Kircher (1602–80) was born at Geisa, became professor at Würzburg, and went to the Jesuit College, Rome, in 1631. He wrote 35 volumes on widely ranging subjects, the most interesting from our viewpoint being *Oedipus Aegyptiacus* (3 vols, Rome, 1652–53) dealing with Egyptian alchemy and quoting Arabic texts; and *Mundus subterraneus* (2 vols, Amsterdam, 1665), which contains much chemical material.

Though he finally settled in Holland, Johann Rudolf Glauber (1604–70) was born at Karlstadt, and was largely self-taught from books. He prepared chemicals for sale, advertising his preparations in his numerous writings, which were collected together as *Opera chymica* (2 vols, Frankfurt, 1658) with *Continuatio operum chymicorum* (Frankfurt, 1659; English translation by Christopher Packe, 1689). Glauber was a firm believer in alchemy but a great practical chemist, inventing furnaces and stills, and conducting experiments on acids, alkalis and salts. It has been suggested that Johann Joachim Becher (1635–82) was more successful with theories than with experiments. Born in Speyer, he was largely self-taught but became professor of medicine at Mainz (1663) and physician to the Elector of Munich (1664). Also a believer in alchemy, Becher was the author of numerous publications, many of which went into several editions. They include *Actorum Laboratorii Chymici Monacensis, seu Physicae subterraneae* (Frankfurt, 1667), to which three supplements with various lengthy titles were published between 1671 and 1680; *Institutiones chimicae prodromae* (Frankfurt, 1664); *Chymischer Glücks-Hafen* (Frankfurt, 1682); *Opuscula chymica rariora* (Nuremburg and Altdorf, 1719); and *Chymischer Rosengarten* (Nuremburg, [1717]).

John Mayow (1641–79), a native of Cornwall, graduated in both law and medicine. His works on rickets and the circulation are best known, but his chemical experiments were significant and are recorded in *Tractatus quinque medico-physici* (Oxford, 1674).

An eminent medical man responsible for the advancement of the phlogiston theory, which had originated with the Greeks, was Georg Ernst Stahl (1660–1734), of Ansbach, Bavaria. Stahl was one of the outstanding chemists of his period, and his extensive writings include

Theoria medica vera, physiologiam et pathologiam (Halle, 1708), his collected medical writings in which he belittled the value of anatomy, physics and chemistry in medicine; *Fragmentorum aetiologiae physiologico-chymicae ex indagatione sensu-rationali* (Jena, 1683); *Observationes chymico-physico-medicae curiosae, mensibus singulis* (i-xi, Frankfurt and Leipzig, 1697–98); *Observationes physico-chymico-medicae curiosae* (Halle, 1709); *Chymica rationalis et experimentalis*; *oder Gründliche der Natur und Vernunfft gemässe* (Leipzig, 1720); and *Fundamenta chymiae dogmaticae & experimentalis* (Nuremburg, 1723).

Several outstanding chemists also contributed usefully to other branches of knowledge, and some conducted their experimental work as a sideline to their main professional careers, which included medicine and the law. Hermann Boerhaave (1668–1738) was the greatest teacher of medicine during the eighteenth century, but he was also a botanist and occupied the chair of chemistry at Leyden University. His treatise on chemistry was pirated in 1724, but the genuine version, signed by the author on the reverse of every title-page, was published in 1732 as *Elementa chemiae* (2 vols, Leyden). It was translated into English, French, German and Latin, and went into numerous editions. Pierre Joseph Macquer (1718–84) also graduated in medicine, but became professor of chemistry at the Jardin du Roi. He replaced many old names by systematic ones, and was an admirer of Stahl. Macquer's numerous writings include *Elémens de chymie théorique* (Paris, 1749) and *Élémens de chymie pratique* (2 vols, Paris, 1751; English translation by Andrew Reid, 1758). With Antoine Baumé he published *Plan d'un cours de chymie expérimentale et raisonnée avec un discours historique sur la chymie* (Paris, 1757), but Macquer's most significant work was undoubtedly his *Dictionnaire de chymie* (2 vols, Paris, 1766; originally published anonymously), which went into numerous editions and translations.

Joseph Black (1728–99) graduated M.D. at Edinburgh with a thesis which he later expanded into *Experiments upon magnesia alba, quicklime and other alkaline substances* (Edinburgh, 1777 and 1782). Black became professor of chemistry at Edinburgh, and was the pioneer of quantitative analysis and pneumatic chemistry. He also wrote *Directions for preparing aerated mineral waters* (Edinburgh, 1787), and his lectures were posthumously published as *Lectures on the elements of chemistry, delivered in the University of Edinburgh* (2 vols, Edinburgh, London, 1803). It has been said of Joseph Priestley (1733–1804) that his defective knowledge of basic chemistry led him into many errors, yet among his numerous discoveries were oxygen, hydrogen chloride, ammonia, sulphur dioxide and carbon monoxide. The first of his numerous scientific works was *The history and present state of electricity* (1767), but the most significant was *Experiments and observations on different kinds of air* (3 vols, 1774–77). This was continued in *Experiments and observations relating to various branches of natural philosophy* (3 vols, 1779–86). Priestley also wrote *Experiments and observations relating to the analysis of atmospherical air* (Philadelphia, 1796); *Considerations on the doctrine of phlogiston* (Philadelphia, 1796); and *The doctrine of phlogiston established* (Northumberland, USA, 1800).

Richard Kirwan (1733–1812) was born in Co. Galway and educated in France, but after studying law turned to science, becoming President of the Royal Irish Academy. He corrected several errors made by Priestley, and wrote papers on the combining proportions of acids and alkalis, but has been described as brilliant but eccentric. His most important contributions to the literature of chemistry were *An essay on phlogiston and the constitution of acids* (1787) and *An essay on the analysis of mineral waters* (1799).

Tobern Olof Bergman (1735–84) studied law, mathematics, physics, chemistry, botany and entomology at Uppsala University, where he became professor of chemistry and pharmacy. He contributed two essays on the history of chemistry, introduced numerous terms into chemical nomenclature and was the author of many printed works, mostly on inorganic chemistry. These were collected together and published as *Opuscula physica et chemica* (3 vols, Stockholm [etc.], 1779–83, and 6 vols, Leipzig, 1788–90). English translations include *Physical and chemical essays* (2 vols, 1784) and *An essay on the general usefulness of chemistry* (1784). Another lawyer, Louis Bernard Guyton de Morveau (1737–1816), later studied chemistry and lectured on the subject at Dijon, his birthplace. These lectures were published with those of Hugues Maret and Jean Francois Durande as *Élémens de chymie, théoretique et pratique* (3 vols, Dijon, 1777–78). Guyton de Morveau established the first French soda factory, ran a glass factory and conducted extensive research on all branches of chemistry. His publications include *Digressions académiques ou Essais sur quelques sujets de physique, de chimie & á'histoire naturelle* (Dijon, 1762) and *Mémoire sur les denominations chimiques* (Dijon, 1782), an important memoir on nomenclature.

The first discoverer of oxygen, Carl Wilhelm Scheele (1742–86), became an apothecary's assistant, but was encouraged to study and investigate. Scheele made numerous contributions to chemistry, and discovered many gases and acids. His writings in journals were collected and published as *Sämmtliche physische und chemische Werke* (2 vols, Berlin, 1793), which were translated into Latin, French and English (1786) and as *Collected Papers*, by L. Dobbin (1931), the better English translation.

Antoine Laurent Lavoisier (1743–94) is probably the best-known French chemist and was the author of 60 papers in the *Mémoires* of the Académie des Sciences between 1768 and 1790, and numerous

papers in other periodicals. His writings published in book form include
Opuscules physiques et chimiques (only Vol. 1 published, Paris, 1774);
*Méthode de nomenclature chimique, proposée par MM. de Morveau,
Lavoisier, Bertholet* [sic] *et de Fourcroy* (Paris, 1787); *Traité élémentaire de chimie* (2 vols, Paris, 1789); and the incomplete posthumously
published *Mémoires de chimie* (2 vols, Paris, 1805). His collected works
were published as *Oeuvres de Lavoisier* (6 vols, Paris, 1862–93), with a
seventh containing his correspondence (1955).

An apothecary who studied chemistry from books, and later ran a
pharmacy, Martin Heinrich Klaproth (1743–1817) became the first
professor of chemistry at Berlin University. His discoveries include the
elements uranium and zirconium, and his many publications were
largely devoted to analytical chemistry. They include *Beiträge zur
chemischen Kenntniss der Mineralkörper* (6 vols, Posen, Berlin, 1795–
1810) and (with F. B. Wolff) *Chemisches Wörterbuch* (5 vols, Berlin,
1807–10, with supplement, 4 vols, Berlin, 1816–19).

Claude Louis Berthollet (1748–1822) graduated as a physician and
studied under Macquer before becoming professor of chemistry at the
École Polytechnique, Paris. A founder of physical chemistry, Berthollet
discovered hydrocyanic acid and hydrogen sulphide, and was the author
of *Recherches sur les lois de l'affinité* (Paris, 1801), which was published in an enlarged form as *Statique chimique* (2 vols, Paris, 1803).
The originator of modern pathological chemistry, Antoine François de
Fourcroy (1755–1809) also studied medicine, but became professor of
chemistry at the Jardin du Roi, and later also at the École de Medecine.
Among his publications the following are particularly noteworthy:
Leçons élémentaires d'histoire naturelle et de chimie (2 vols, Paris,
1782), with the supplementary *Mémoires et observations de chimie*
(Paris, 1784); *Système des connaissances chimiques* (11 vols, 1801–2);
and *Philosophie chimique* (Paris, 1792).

Jean Antoine Chaptal (1756–1832) was professor of chemistry at
Montpellier before becoming Minister of the Interior in 1800. He pro-
moted the development of agriculture, and wrote *Tableau analytique
du course de chymie* (Montpellier, 1783), *Élémens de chimie* (3 vols,
Paris and Montpellier, 1790), *Chimie appliquée aux arts* (4 vols, Paris,
1807). The name 'stoichiometry' was used by Jeremias Benjamin
Richter (1762–1807) for the combining proportions of substances,
and he wrote *Ueber die neuern Gegenstände der Chymie* (11 parts,
Breslau [etc.], 1791–1802). *Anfangsgründe der Stöchyometrie oder
Messkunst chymischer Elemente* (3 vols [in 4], Breslau, Hirschberg,
1792–94) and *Chemisches Handwörterbuch* (6 vols, Berlin, 1802–5).

Largely self-taught, John Dalton (1766–1844) taught mathematics
and chemistry, propounded an atomic theory of matter, invented a
system of chemical symbols, and discovered the law of expansion of gases
by heat and the law of partial pressures. His main separately published

work was the uncompleted *A new system of chemical philosophy* (Parts i–ii and 2, i, Manchester, London, 1808–27). The atomic theory of William Higgins (1763–1825) preceded that of Dalton, but is considered inferior. Higgins became chemist at Apothecaries' Hall, Dublin, and later to the Irish Linen Board. His books include *A comparative view of the phlogistic and antiphlogistic theories* (1789), *An essay on the theory and practice of bleaching* (Dublin, 1799) and *Experiments and observations on the atomic theory, and electrical phenomena* (Dublin, 1814).

Nineteenth century chemistry abounds with the names of great chemists, and selection for mention is difficult. Michael Faraday (1791–1867) contributed mainly to physics, but he held the chair of chemistry at the Royal Institution, discovered benzol and butylene, and was the author of *Chemical manipulations* (1827); *Experimental researches in chemistry and physics* (1859), a collection of his articles originally published in journals; and *A course of six lectures on the chemical history of the candle* (1861). Friedrich Accum (1769–1838), a Westphalian domiciled in England, wrote numerous books, including *System of theoretical and practical chemistry* (2 vols, 1803), *Manual of a course of lectures on experimental chemistry and on mineralogy* (1810), *A practical essay on chemical reagents or tests* (1816) and *A treatise on adulteration of food and culinary poisons* (1820). The founder of one of the earliest chemical laboratories in Great Britain, Thomas Thomson (1773–1852), was Regius professor of chemistry at Glasgow. He wrote an outstanding textbook which went into numerous editions and translations, *A system of chemistry* (4 vols, Edinburgh, 1802), which was followed by *Elements of chemistry* (1810), *A history of chemistry* (2 vols, 1830–31), *An attempt to establish the first principles of chemistry by experiment* (2 vols, 1825) and *Chemistry of organic bodies. Vegetables* (1838).

Amedeo Avogadro (1776–1856), first professor of mathematical physics at Turin, put forward a hypothesis in his paper 'Essai d'une manière de déterminer les masses rélatives des molécules des corps, [etc.]' (*Journal de Physique,* 73, 58–76, 1811), which, while not accepted for many years, is now regarded as a cornerstone of the theory of gases. His selected works were published as *Opere scelte* (Turin, 1911).

A professor at the Sorbonne, the French chemist Louis Jacques Thenard (1777–1857) discovered hydrogen peroxide and made important investigations in organic chemistry. He wrote *Traité de chimie élémentaire* (Paris, 1813–16) and with Joseph Louis Gay-Lussac (1778–1850) was the joint author of *Recherches physico-chimiques* (2 vols, Paris, 1811). Gay-Lussac discovered cyanogen, conducted research on iodine and made numerous important scientific discoveries.

Sir Humphry Davy (1778–1829) was the inventor of the miner's safety lamp, and conducted research on electrolysis, chlorine and oxides of fluorine. Davy described the anaesthetic properties of nitrous oxide in *Researches chemical and philosophical, chiefly concerning nitrous oxide, or dephlogisticated nitrous air, and its respiration* (1800) and was also the author of *On the safety lamp for coal miners: with some researches on flame* (1818), *Elements of chemical philosophy* (Vol. 1, i, 1812) and *Elements of agricultural chemistry* (1813). His writings were eventually issued as *Collected works* (9 vols, 1839–40).

Jöns Jacob Berzelius (1779–1848) was professor of chemistry at the Chirurgico-Medical Institute, Stockholm, and discovered several elements. He also determined the atomic and molecular weights of various substances, and wrote *Lärbok i kemien* (6 parts, Stockholm, 1808–30) a textbook which went into numerous editions and translations.

Three German chemists of this period are noteworthy for their contributions: Leopold Gmelin (1788–1853), professor of chemistry at Heidelberg, who founded the monumental *Handbuch der theoretischen Chemie* (Frankfurt, 1817–19, translated into English in nineteen volumes, 1848–72) (see Chapter 9); Friedrich Wöhler (1800–82), author of *Grundriss der anorganischen Chemie* (Berlin, 1831) and *Grundriss der organischen Chemie* (Berlin, 1840); and Justus von Liebig (1803–73), who became professor of chemistry at Giessen (1825) and Munich (1852). Liebig was a pioneer in agricultural and physiological chemistry, and was eminent as editor of *Annalen der Chemie* and as a teacher. His books include *Die organische Chemie in ihrer Anwendung auf Agricultur und Physiologie* (Braunschweig, 1840), *Die Thierchemie oder die organische Chemie in ihrer Anwendung auf Physiologie und Pathologie* (Braunschweig, 1842), *Handbuch der organischen Chemie mit Rücksicht auf Pharmacie* (Heidelberg, 1843), *Chemische Briefe* (Heidelberg, 1844), *Die Grundsätze der Agricultur-Chemie* (1855) and a collection of his articles from journals, *Reden und Abhandlungen* (Leipzig, [etc.], 1874).

Jean Baptiste André Dumas (1800–84) was one of the most outstanding of the French chemists of the period, and conducted research on the atomic theory, vapour densities and physiological chemistry. He wrote *Mémoire sur les substances vegetales qui se rapprochement due camphre, et sur quelques huiles essentielles* (Paris, 1832), *Leçons sur la philosophie chimique* (Paris, 1837) and *Traité de chimie appliquée aux arts* (8 vols, Paris, Brussels, 1828–46). August Laurent (1807–53) was assistant to Dumas, and conducted extensive research in organic chemistry and on atomic weights. He wrote *Méthode de chimie* (Paris, 1854) and, with Charles Frederic Gerhardt (1816–56), *Comptes rendus mensuels des travaux chimiques. Années* 1845–51 (7 vols, Paris,

1846–59). Gerhardt also wrote *Précis de chimie organique* (2 vols, Paris, 1844–45) and *Traité de chimie organique* (Paris, 1853–56). The Bunsen battery, burner and cell are all named after Robert Wilhelm Bunsen (1811–99), who was professor of chemistry at Heidelberg. He discovered caesium and rubidium and was the author of *Gasometrische Methoden* (Braunschweig, 1857) and *Chemische Analyse durch Spectralbeobachtungen* (Vienna, 1860). His collected works were published as *Gesammelte Abhandlungen* (3 vols, Leipzig, 1904). Hermann Kopp (1817–92) also became professor of chemistry at Heidelberg, and conducted research in physical chemistry, in addition to his historical work. He wrote *Geschichte der Chemie* (4 vols, Braunschweig, 1843–47), *Beiträge zur Geschichte der Chemie* (Braunschweig, 1869–75), *Die Entwicklung der Chemie in die neueren Zeit* (Munich, 1873) and *Die Alchemie in älterer und neuerer Zeit* (Heidelberg, 1886). Both Kopp and August Wilhelm von Hofmann (1818–92) studied under Liebig, and Hofmann became prominent in the development of the dye industry. He was the author of *On mauve and magenta* (1862) and *Introduction to modern chemistry, experimental and theoretic* (1865).

The founder of modern bacteriology was also one of France's greatest chemists, and achieved world-wide recognition for his investigations on silk-worm disease, chicken cholera, hydrophobia and diseases of wine. Louis Pasteur (1822–95) wrote *Études sur le vin* (Paris, 1866) and *Études sur le vinaigre* (Paris, 1868), among other books, and his writings were collected together and published in two volumes (Paris, 1922) and as *Oeuvres* (7 vols, Paris, 1923–39). Pierre Eugène Marcellin Berthelot (1827–1907) was professor of organic chemistry at the Collége de France, and wrote extensively on the history of chemistry. A pioneer in thermochemistry, Berthelot conducted experiments on the synthesis of organic compounds and on explosives. He wrote *Chimie organique fondée sur la synthèse* (Paris, 1860), *Sur la force de la poudre et des matières explosives* (Paris, 1871), *Essai de mécanique chimique* (2 vols, 1879). *Thermochimie* (Paris, 1897) and *Les carbures d'hydrogene* (3 vols, 1901).

The benzene hexagon was but one of Friedrich August Kekulé's (1829–96) many contributions to the periodical literature of organic chemistry. His *Lehrbuch der organischen Chemie, oder der Chemie der Kohlenstoffverbindungen* (Vols 1–3, 4, i, Erlangen, Stuttgart, 1861–87) was not completed.

A pupil under Hofmann, Sir William Crookes (1832–1919) concerned himself with radiation and spectroscopy, discovering thallium. He wrote *Selected methods in chemical analysis* (1871), *Practical handbook of dyeing and calico printing* (1874) and *Diamonds* (1909).

A professor of organic chemistry at Manchester, Carl Schorlemmer* (1834–92) experimented with the paraffin hydrocarbons, alcohols, aurin and suberone. He wrote *A manual of the chemistry of the carbon compounds* (1874), *Der Ursprung der organischen Chemie* (Braunschweig, 1889) and, with Sir Henry Enfield Roscoe (1833–1915), *A treatise on chemistry* (2 vols, 1878–1911).

The eminent Russian chemist Dmitri Ivanovich Mendeléeff (1834–1907) investigated the properties of solutions and the thermal expansion of fluids, and formulated his periodic law in 1868–69. His textbook on chemistry, first published at St Petersburg, 1868–70, was translated into English from the 5th Russian edition as *The principles of chemistry* (1891) and he wrote *La loi périodique des éléments chimiques* (1879). Mendeléeff's writings were collected together as *Sochineniya* (25 vols, Leningrad, 1934–54).

Sir Thomas Edward Thorpe (1845–1925) conducted research on the derivatives of fluorine, phosphorus and the paraffin hydrocarbons, but is mainly known as a historian of chemistry. He was the author of *A history of chemistry* (2 vols, 1909–10), *Essays in historical chemistry* (3rd edn, 1911) and *Alcoholometric tables* (1915), and founded Thorpe's *Dictionary of Applied Chemistry* (3 vols, 1890–93) (see Chapter 7). Sir William Ramsay (1852–1916), who discovered the inert gases, was another historian of his subject. Ramsay was the author of *Gases of the atmosphere: the history of their discovery* (1896), *Modern chemistry* (1900), *Essays biographical and chemical* (1908), *Elements and electrons* (1912) and *The life and letters of Joseph Black* (1918).

Jacobus Henricus van't Hoff (1852–1911), of Amsterdam, was a pupil of Kekulé and conducted research on the stereochemistry of carbon, the osmotic theory of solutions and other aspects of physical chemistry. He was the author of *Études de dynamique chimique* (Amsterdam, 1884) and *Dix années dans l'histoire d'une théorie: Deuxième édition de 'La chimie dans l'espace'* (Rotterdam, 1887). Another physical chemist, Wilhlem Ostwald (1853–1932), conducted research in electrochemistry, kinetics and catalysis. He wrote *Lehrbuch der allgemeine Chemie* (2 vols, Leipzig, 1885–87), *Grundlinien der anorganischen Chemie* (5th edn, Leipzig, 1922), *Die wissenschaftlichen Grundlagen der analytischen Chemie* (Leipzig, 1894) and the first guide to the chemical literature (1919). Edmund O. von Lippman (1857–1940) conducted research on the sugars, and was also a historian of chemistry. He was the author of *Der Zucker, seine Derivate und sein Nachweis* (Vienna, 1878), *Die Zuckerarten und ihre Derivate* (Braunschweig, 1882), *Die Chemie der Zuckerarten* (Braunschweig, 1890),

*A friend of Marx and Engels, he is now the 'patron saint' of chemistry in East Germany.

Geschichte des Zuckers (Leipzig, 1890) and *Beiträge zur Geschichte der Naturwissenschaften und Technik* (Berlin, 1923). The Nernst equation was formulated by Hermann Walther Nernst (1864–1941), who was also responsible for the Third Law of Thermodynamics and for the Nernst incandescent electric lamp. Nernst was the author of *Theoretische Chemie* (Stuttgart, 1893), *Die theoretischen und experimentellen Grundlagen des neuen Warmesatzes* (1918), and various other works which were translated into English.

These represent some of the outstanding contributions to the literature of chemistry, a subject that has grown out of all proportion during this century. Chemists specializing within different fields find it difficult to keep pace with the output of literature devoted to comparatively minute branches of the subject, and chemists no longer speak a common language. The history of chemistry is still being written, but it will never again be adequately documented by a single historian covering the entire field.

APPENDIX: SELECTED LIST OF BOOKS ON THE HISTORY OF CHEMISTRY AND BIOGRAPHIES OF CHEMISTS

No attempt has been made to list historical and biographical articles in journals, but such articles can be traced in *A History of Chemistry*, by J. R. Partington, and in *Scientific Books, Libraries and Collectors*, by the authors of this chapter. A short list of such periodicals appears at the end of this Appendix. Bibliographies of the writings of individuals are included, and place of publication is London unless otherwise indicated. Where substantial modern biographies have superseded older works, the latter have been omitted.

General histories of science

Crombie, A. C. *Augustine to Galileo*, 2nd edn, 2 vols, 1961 (published also under the title *Medieval and Modern Science*, 2 vols, New York)

Dingle, H., Ed. *A Century of Science, 1851–1951*, 1951

Farrington, B. *Greek Science: its Meaning for us*, rev. edn, 1953

Forbes, J. R. and Dijksterhuis, E. J. *A History of Science and Technology*, 2 vols, Harmondsworth, 1963

Hall, A. R. *The Scientific Revolution, 1500–1800. The Formation of the Modern Scientific Attitude*, 2nd edn, 1962

Nasr, S. H. *Science and Civilisation in Islam*, 1968

Needham, J. *Science and Civilisation in China*, 7 vols. Cambridge, 1954– . Vol. 5, *Chemistry and Chemical Technology*, Part II, 1974

Pledge, H. T. *Science since 1500: a Short History of Mathematics, Physics, Chemistry, Biology*, 2nd edn, 1966

Rider, K. J. *The History of Science and Technology: a Select Bibliography for Students*, 2nd edn, 1970. Library Association Special Subject List, No. 48

Sarton, G. *Introduction to the History of Science*, 3 vols [in 5], Baltimore, 1927–48

Sarton, G. *Horus. A Guide to the History of Science. A First Guide for the Study of the History of Science, with Introductory Essays on Science and Tradition*, Waltham, Mass., 1952

Sarton, G. *A History of Science: Ancient Science through the Golden Age of Greece*, 1953

Sarton, G. *A History of Science (II): Hellenistic Science and Culture in the Last Three Centuries B.C.*, Cambridge, Mass., London, 1959

Singer, C. *et al.*, Eds. *A History of Technology*, 5 vols, Oxford, 1954–58

Singer, C. *A Short History of Scientific Ideas to 1900*, Oxford, 1959

Taton, R., Ed. *A General History of the Sciences*, translated by A. J. Pomerans, 4 vols, 1964–66

Thornton, J. L. and Tully, R. I. J. *Scientific Books, Libraries and Collectors. A Study of Bibliography and the Book Trade in Relation to Science*, 3rd edn, 1971; reprinted 1975 (*supplement 1969–75*, 1978)

Wolf, A. *A History of Science, Technology and Philosophy in the 16th & 17th Centuries*, new edn prepared by D. McKie [etc.], 1950 [1951]

Wolf, A. *A History of Science, Technology and Philosophy in the 18th Century*, 2nd edn revised by D. McKie [etc.], 1952

Histories of chemistry including alchemy

NOTE: A useful general survey of the histories of chemistry up to 1920 is provided by Farber (1965).

Asimov, I. *A Short History of Chemistry*, 2nd edn, 1972

Berry, A. J. *From Classical to Modern Chemistry: Some Historical Sketches*, Cambridge, 1954

Browne, C. A. and Weeks, M. E. *A History of the American Chemical Society: Seventy-five Eventful Years*, Washington, DC, 1952

Clow, A. and N. *The Chemical Revolution*, 1952

Crosland, M. P. *Historical Studies in the Language of Chemistry*, 1962

Debus, A. G. *The English Paracelsians*, 1965

Farber. E., Ed. *Milestones of Modern Chemistry*, New York, 1966

Farber, E. *The Evolution of Chemistry* [etc.], 2nd edn, New York, 1969

Findlay, A. *A Hundred Years of Chemistry*, 3rd edn revised by T. I. Williams, 1965

Fox, R. *The Caloric Theory of Gases from Lavoisier to Regnault*, Oxford, 1971

Hill, C. R. *Museum of the History of Science. Catalogue 1: Chemical Apparatus*, 1971

Holmyard, E. J. *Alchemy*, Harmondsworth, 1957; reprinted 1968

Hutin, S. *A History of Alchemy*, translated by T. Alferoff, New York, 1963

Ihde, A. J. *The Development of Modern Chemistry*, New York [etc.], 1964

Jaffe, B. *Crucibles: the Story of Chemistry from Alchemy to Nuclear Fission*, rev. edn, New York, 1948

Leicester, H. M. *The Historical Background to Chemistry*, New York, 1956

Moore, F. J. *A History of Chemistry*, revision prepared by W. T. Hall, 3rd edn, New York, 1939

Multhauf, R. P. *The Origins of Chemistry*, 1966

Partington, J. R. *A Short History of Chemistry* [*etc.*] , 3rd edn, 1957

Partington, J. R. *A History of Chemistry*, 4 vols, 1961–70 (Vol. 1, Part 1, 1970; Vol. 2, 1961; Vol. 3, 1962; Vol. 4, 1964)

Poynter, F. N. L., Ed. *Chemistry in the Service of Medicine*, 1963

Read, J. *Prelude to Chemistry; an Outline of Alchemy, its Literature and Relationships*, 2nd edn, 1939; reprinted 1961

Read, J. *Through Alchemy to Chemistry; a Procession of Ideas & Personalities*, 1957

Routh, J. I. *20th Century Chemistry*, 3rd edn, Philadelphia, 1963

Stillman, J. M. *The Story of Alchemy and Early Chemistry*, New York, 1960 (Dover Books); previously with title *The Story of Early Chemistry*, New York, 1924

Szabadváry, F. *History of Analytical Chemistry*, translated by G. Svehla, 1966

Weeks, M. E. *Discovery of the Elements*, 7th edn, 1968

Biographies and bibliographies of chemists

In addition to entries in scientific reference books, material on individual scientists will also be found in general works such as *Dictionary of National Biography*; *Dictionary of American Biography*; *Nouvelle Biographie Universelle (Générale)* [*etc.*]

Collective

Aykroyd, W. R. *Three Philosophers (Lavoisier, Priestley, Cavendish)*, 2nd edn, 1936

Asimov, I. *Asimov's Biographical Encyclopaedia of Science and Technology* [*etc.*] , New York, 1964

Bugge, G., Ed. *Das Buch der grossen Chemiker*, 2 vols, Berlin, 1929; reprinted 1955

Crowther, J. G. *Famous American Men of Science*, 1937

Farber, E., Ed. *Great Chemists*, 1961

Farber, E., Ed. *Nobel Prize Winners in Chemistry, 1901–1961*, rev. edn, 1963

Findlay, A. and Mills, W. H., Eds. *British Chemists*, 1947

Gillespie, C. C., Ed. *Dictionary of Scientific Biography*, 14 vols + index, 1970–76

Hartley, Sir H. *Studies in the History of Chemistry*, Oxford, 1971

Holmyard, E. J. *The Great Chemists . . .*, 3rd edn, 1929

Howard, A. V. *Chambers's Dictionary of Scientists*, 1951

Nobel Foundation. *Nobel Lectures, including Presentation Speeches and Laureates, and Biographies. Chemistry*, 3 vols, Amsterdam, London, 1964–66

Poggendorff, J. C. *Biographisch-literarisches Handwörterbuch zur Geschichte der exacten Wissenschaften*, Berlin, 1863– [reprinting 1965–]

Prandtl, W. *Deutsche Chemiker in ersten Hälfte des 19 Jahrhunderts*, Weinheim, 1956

Smith, H. M. *Torchbearers of Chemistry: . . . With Bibliographies of Biographies, by R. E. Oesper*, New York, 1949

Tilden, W. *Famous Chemists: the Men and their Work*, 1921

Individual

AVOGADRO

Meldrum, A. N. *Avogadro and Dalton. The Standing in Chemistry of their Hypothesis*, Edinburgh, 1904

BERGMAN

Moström, Borgitta. *Tobern Bergman. A Bibliography of his Works*, Stockholm, 1957

BERTHELOT

Boutaric, A. *Marcellin Berthelot*, Paris, 1927

Velluz, L. *Vie de Berthelot*, Paris, 1964

BERZELIUS

Holmberg, A. *Bibliographie de J. J. Berzelius [etc.]*, 2 vols and suppls, 1933– (Vol. 1, 1933, Suppl. 1, 1936, Suppl. 2, 1953; Vol. 2, 1936, Suppl. 1, 1953)

Jorpes, J. E. *Jac. Berzelius, his Life and Work*, Stockholm, 1966

Söderbaum, H. G. *Jac. Berzelius Levnadsteckning*, 3 vols, Uppsala, 1929–31

BLACK

Donovan, A. L. *Philosophical Chemistry in the Scottish Enlightenment: the Doctrines and Discoveries of William Cullen and Joseph Black*, Edinburgh, 1976

Ramsay, Sir W. *The Life and Letters of Joseph Black [etc.]*, 1918

BOERHAAVE

Lindeboom, G. A. *Bibliographia Boerhaaviana [etc.]*, Leyden, 1959

Lindeboom, G. A. *Herman Boerhaave. The Man and his Work*, 1968

BOYLE

Boas, M. *Robert Boyle and Seventeenth-Century Chemistry*, Cambridge, 1958

Fulton, J. F. *A Bibliography of the Honourable Robert Boyle, F.R.S.*, 2nd edn, Oxford, 1961

Maddison, R. E. W. *The Life of the Honourable Robert Boyle, F.R.S.*, 1969

BUNSEN

Lockemann, G. *Robert Wilhelm Bunsen. Lebensbild eines deutschen Naturforscher*, Stuttgart, 1949

CHAPTAL

Pigeire, J. *La Vie et l'Oeuvre de Chaptal*, Paris, 1932

CROOKES

Fournier d'Albe, E. E. *The Life of Sir William Crookes*, 1923

DALTON

Brockbank, E. M. *John Dalton: Some Unpublished Letters of Personal and Scientific Interest [etc.]*, Manchester, 1944

Cardwell, D. S. L., Ed. *John Dalton and the Progress of Science [etc.]*, Manchester, New York, 1968

Greenaway, F. *John Dalton and the Atom*, 1966

Patterson, E. *John Dalton and the Atomic Theory*, 1970

Smyth, A. L. *John Dalton 1766–1844. A Bibliography of Works by and about him*, Manchester, 1966

Thackray, A. *John Dalton. Critical Assessments of his Life and Science*, Cambridge, Mass., London, 1972

DAVY

Davy, J. *Memoirs of the Life of Sir Humphry Davy, Bart. By his Brother*, 2 vols, 1836

Fullmer, June Z. *Sir Humphry Davy's Published Works*, 1969

Gregory, J. C. *The Scientific Achievements of Sir Humphry Davy*, Oxford, 1930

Hartley, Sir H. *Humphry Davy*, 1966

Treneer, H. *The Mercurial Chemist: a Life of Sir Humphry Davy*, 1963
DUMAS
Dumas, J.-B., fils. *La Vie de J.-B. Dumas*, Paris, 1924
Maindron, E. *L'Oeuvre de J. B. Dumas*, Paris, 1866
FARADAY
Agassi, J. *Faraday as a Natural Philosopher*, Chicago, London, 1972
Faraday, M. *The Selected Correspondence*, [*etc.*], 2 vols, 1971
Jeffreys, A. E. *Michael Faraday: a List of his Lectures and Published Writings,*
    [*etc.*], 1960
Williams, L. P. *Michael Faraday. A Biography*, 1965
FOURCROY
Kersaint, G. *Antoine François de Fourcroy, sa Vie et son Oeuvre*, Paris, 1966
Smeaton, W. A. *Fourcroy: Chemist and Revolutionary, 1755–1809*, Cambridge,
    1962
GAY-LUSSAC
Blanc, E. and Delhoume, L. *La Vie émouvante et noble de Gay-Lussac*, Paris,
    1950
GMELIN
Pietsch, E. *Leopold Gmelin*, Berlin, 1938
GUYTON DE MORVEAU
Bouchard, G. *Guyton-Morveau, Chimiste et Conventionelle (1737–1816)*, Paris,
    1938
VAN HELMONT
Nève de Mévergnies, P. *Jean Baptiste van Helmont, Philosophe par le Feu*, Liege,
    1935
Waele, H. de. *J. B. van Helmont*, Brussels, 1947
HIGGINS
Wheeler, T. S. and Partington, J. R. *The Life and Work of William Higgins, Chemist
    (1763–1825)*, [*etc.*], Oxford [etc.], 1960
HOFF, J. H. VAN'T
Cohen, E. J. *Jacobus Henricus van't Hoff, sein Leben und Werken*, Leipzig, 1912
KEKULÉ
Anschütz, R. *August Kekulé. I. Leben und Werken; II. Abhandlungen*, [*etc.*],
    Berlin, 1929
KLAPROTH
Dann, G. E. *Martin Heinrich Klaproth (1743–1817) . . . Sein Weg und seine
    Leistung*, Berlin, 1958
LAVOISIER
Duveen, D. I. and Klickstein, H. S. *A Bibliography of the Works of Antoine
    Laurent Lavoisier, 1743–1794*, [*etc.*], 1954; supplement by D. I. Duveen,
    1965
Grimaux, E. *Lavoisier 1743–1795* [sic; 1794], *d'après sa Correspondence, ses
    Manuscrits, ses Papiers de Famille et d'autre Documents inédits*, 3e édn, Paris,
    1899
Guerlac, H. *Lavoisier – the Crucial Year: the Background and Origin of his First
    Experiments on Combustion in 1772*, Ithaca, N.Y., 1961
McKie, D. *Antoine Lavoisier, the Father of Modern Chemistry*, 1935
McKie, D. *Antoine Lavoisier, Scientist, Economist, Social Reformer*, 1952
Velluz, L. *Vie de Lavoisier*, Paris, 1966
LIEBIG
Dechend, H. von. *Justus von Liebig in eigenen Zeugnissen und solchen seiner
    Zeitgenossen*, Weinheim, 1963
Paolini, C. *Justus von Liebig: eine Bibliographie sämtlicher Veröffentlichungen*,
    Heidelberg, 1968

Volhard, J. *Justus von Liebig*, 2 vols, Leipzig, 1909
MACQUER
Coleby, L. J. M. *The Chemical Studies of P. J. Macquer*, 1938
MENDELÉEFF
Pisarzhevsky, D. *Ivanovich Mendeleev. His Life and Work*, Moscow, 1959
Posin, D. Q. *Mendeleyev. The Story of a Great Scientist*, New York, 1948
PARACELSUS
Kerner, D. *Paracelsus, Leben und Werk*, Stuttgart, 1965
Pachter, H. M. *Paracelsus: Magic into Science*, New York, 1951
Pagel, W. *Paracelsus. An Introduction to Philosophical Medicine in the Era of the Renaissance*, Basle, New York, 1958
Sudhoff, K. *Paracelsus. Ein deutsches Lebensbild aus dem Tagen der Renaissance*, Leipzig, 1936
PASTEUR
Cuny, H. *Louis Pasteur: the Man and his Theories*, 1965
Dubos, R. J. *Louis Pasteur, Free Lance of Science*, 1951
Nicolle, J. *Louis Pasteur: a Master of Scientific Enquiry*, 1961
Vallery-Radot, R. *The Life of Pasteur . . . Translated from the French by Mrs. R. L. Devonshire*, [etc.], 2 vols, 1902
PRIESTLEY
Crook, R. E. *A Bibliography of Joseph Priestley 1733–1804*, 1966
Gibbs, F. W. *Joseph Priestley*, Edinburgh, 1965
Holt, A. *A Life of Joseph Priestley*, [etc.], 1931
Priestley, J. *Memoirs of Dr Joseph Priestley, to the Year 1795, Written by himself*, [etc.], 2 vols, 1806–07
RAMSAY
Tilden, Sir W. A. *Sir William Ramsay, K.C.B., F.R.S.: Memorials of his Life and Work*, 1918
Travers, M. W. *A Life of Sir William Ramsay, K.C.B., F.R.S.*, 1956
ROSCOE
Thorpe, T. E. *The Rt. Hon. Sir H. E. Roscoe, a Biographical Sketch*, 1916
SCHEELE
Boklund, U., Ed. *Carl Wilhelm Scheele. His Work and Life*, 8 vols, Stockholm, 1969–   [in progress]
WÖHLER
Valentin, J. *Friedrich Wöhler*, Stuttgart, 1949

Periodicals

A history of early chemical periodicals *per se* has been given by Yagello (1968).

1. *Ambix*, 1937–
2. *Annals of Science*, 1936–
3. *Archeion: Archivio di Storia della Scienza*, 1927–47 [*continuation of* 6; *continued as* 5]
4. *Archive for History of Exact Sciences*, 1960–
5. *Archives internationales d'Histoires des Sciences*, 1947–   [*continuation of* 3]
6. *Archivio di Storia della Scienza*, 1919–27 [*continued as* 3]
7. *Biographical Memoirs of Fellows of the Royal Society*, 1955–   [*continuation of* 22]

 8. *British Journal for the History of Science*, 1962–
 9. *British Journal for the Philosophy of Science*, 1950–
10. *Chemistry and Industry*, 1923–
11. *Chymia: Annual Studies in the History of Chemistry, 1948–68* [*continued in* 14]
12. *Discovery*, 1920–66
13. *Endeavour*, 1942–
14. *Historical Studies in the Physical Sciences*, 1969–
15. *History of Science*, 1962–
16. *Isis*, 1912–
17. *Journal of Chemical Education*, 1924–
18. *Journal of the History of Ideas*, 1940–
19. *Monographien aus der Geschichte der Chemie*, 1897–1904
20. *Nature*, 1869–
21. *New Scientist*, 1956–
22. *Obituary Notices of Fellows of the Royal Society*, 1932–54 [*continued as* 7]
23. *Osiris*, 1936–68
24. *School Science Review*, 1919–
25. *Science Progress*, 1906–
26. *Scientific American*, 1846–

NOTE: Publishers' names have been deliberately omitted from this appendix (cf. Exercise 4 (4)).

REFERENCES

Farber, E. (1965). *J. Chem. Educ.*, **42**, 120
Yagello, V. E. (1968). *J. Chem. Educ.*, **45**, 426

18

Practical use of the chemical literature

R. T. Bottle

The introductory chapter indicates the structure of this book, showing how the various chapters serve to answer certain problems. A number of cross-references have, of course, been necessary and this chapter will contain many more. It is thus intended to be a broad index to the book's use. There is, however, little point in parading a vast number of titles with annotations before the reader and saying much treasure awaits him in the literature. The key is practice; the more one uses the literature, the more information one will gather from it. Remember that information is too valuable a crop to be picked in an inefficient and wasteful manner; some exercises have been included in Appendix 1 so that the novice can gain experience in using the literature. These exercises are those which have been used in several universities' postgraduate courses over the past two decades. Others will be found in Mellon's *Chemical Publications* (q.v.). Some will involve long searches, others will require only short searches. (The term *search* as understood by literature specialists is really equivalent to *use*, since they may almost describe looking up a melting point in the *Rubber Handbook* as a search!)

The ensuing sections of this chapter are, in effect, the various purposes for which the literature is searched (apart from acquiring background information, as is described in Chapter 7). These divisions are, at best, somewhat arbitrary, and there is considerable overlap of one with another depending on the depth of knowledge required.

QUICK REFERENCE

This is probably the purpose for which the literature is most often used. Many of the necessary books should be found in, or adjacent to, any well-equipped laboratory (although it will probably keep only those testing standards, etc., which are indispensable to its routine work). Undergraduate practical manuals will, of course, solve many of the simpler queries in the categories mentioned below.

Physical properties such as melting point, refractive index, solubility, and so on, are frequently required and can usually be found in the ubiquitous *Handbook of Chemistry and Physics* (Chemical Rubber Publishing Co.), currently in its 59th edition, commonly known, even to first-year students, as the 'Rubber Handbook' or the 'Bible'. If, however, one requires more critically compiled data, then reference should be made to the books of tables discussed in Chapter 8. Trade literature available on commercially obtainable chemicals is dealt with in Chapter 15.

Physicochemical techniques are now so widespread and varied that there are few encyclopaedic works in this field and each distinct method is usually described by one or more monographs. A large number, however, are described with ample literature references in A. Weissberger and B. W. Rossiter's *Physical Methods of Chemistry* (5 parts, Interscience, 1971–72). These 10 volumes are part of the *Technique of Chemistry Series*, which is under Weissberger's general editorship. Spectroscopic data sources are discussed in Chapter 8, while spectroscopic and other techniques for structure determination are the subject of a section in Chapter 12. From 1959 to 1962 the *Journal of Chemical Education* carried a most useful basic survey on chemical instrumentation and since 1962 this has been replaced by 'Topics in Chemical Instrumentation', which describes the operation, principles, circuits, etc., of commercially available physicochemical instruments (see the annual indexes). (The same journal is a useful source of advertisements which show what new instruments and apparatus are available in the US.) Applications of the above methods on the plant are dealt with in Considine's *Process Instruments and Controls Handbook* (2nd edn, McGraw-Hill, 1974) and its companion volume, *Handbook of Applied Instrumentation* (D. M. Considine and S. D. Ross, Eds, McGraw-Hill, 1964).

Preparative details for inorganic and organic substances will usually be found in the books mentioned in Chapters 9 and 12, otherwise reference will have to be made to the original publication (located through the formula or chemical substance indexes of *Chemical Abstracts*: see Chapter 5). Heilbron and Bunbury's *Dictionary of Organic Compounds* (Chapter 12) gives the characteristics of some

70 000 organic compounds with literature references to their original preparation. Biochemists will find the series *Biochemical Preparations* (Wiley, 1949–71) a valuable source. Still useful for quick reference in this field are C. Long's *Biochemist's Handbook* (Spon, 1961) and the *Encyclopedia of Biochemistry* (R. J. Williams and E. M. Lansford, Eds, Reinhold, 1967). (The biochemical literature is, however, dealt with in detail in Chapter 14 of *The Use of Biological Literature* or Chapter 7 of *Use of Medical Literature*.) *Laboratory Techniques in Chemistry and Biochemistry* (P. S. Diamond and R. F. Denman, 2nd edn, Butterworths, 1975), although mainly aimed at technicians, contains many useful recipes for a wide range of reagents.

Much of the industrial application of chemicals is concerned with formulation problems. Apart from the advice available from manufacturers and their trade literature (Chapter 15), an invaluable collection of recipes for most conceivable products from adhesives to varnishes, bubble gum to liqueurs, as well as the nostrums covered by the Pharmacopoeias, etc., is contained in H. Bennett's *Chemical Formulary* (20 vols, Chemical Publishing, 1933–74), for which a cumulated index of Vols 1–10 is available. One must, of course, check current legislation to see if they are permitted, as 'Bennett' contains some recipes which are not.

Analysis and testing are frequent operations in laboratories and while for most purposes textbook methods are quite adequate (see Chapters 9 and 12), in certain circumstances the official or standard method must be used. The *BP Codex* (Chapter 7) gives methods to be used and specifications to be met by substances of pharmaceutical interest. In this country the British Standards Institution issues specifications for materials, standard methods of testing, codes of practice, etc., as and when there is a generally recognized need for them. *British Standards* are revised from time to time. For those of chemical interest, consult the appropriate sectional list, e.g. *Chemical engineering*; *Chemicals, fats, oils, scientific apparatus, etc.*; *Non-ferrous metals*; etc. A complete list of all *British Standards* is in the *British Standards Yearbook*, which lists the libraries at home and abroad which house a complete set of standards. The American Standards Association and the American Society for Testing and Materials perform a similar service in the US. The latter's *ASTM Standards* is a most useful encyclopaedia of standards and test methods. Standards, including defence standards, are available from Technical Indexes Ltd in microform or via Prestel. UK, US and West German (DIN) standards for plastics, etc., are discussed briefly in Chapter 13.

Returning to more specifically analytical problems, an authoritative bibliography, originally sponsored by the Society of Public Analysts, is *Official Standardised and Recommended Methods of Analysis* (N. W.

Hanson, Ed., 2nd edn, SAC/Chemical Society, 1973). A useful compilation of procedures for natural products is *Official Methods of Analysis* (12th edn, Association of Official Agricultural Chemists, Washington, DC, 1975). A new edition appears every five years. The *Handbook of Analytical Chemistry* (L. Meites, Ed., McGraw-Hill, 1963) is a useful summary of data and methods but is most valuable as a classified literature key. The 23 volume *Encyclopedia of Industrial Chemical Analysis* (F. D. Snell and C. L. Hilton, Eds, Interscience, 1966–72) is also well documented. *Methods in Chemical and Mineral Microscopy* (E. E. El-Hinnawi, Elsevier, 1966) is a useful laboratory manual in a related field.

Minor theoretical or practical points can be verified by looking in the appropriate monograph or in one of the chemical encyclopaedias or dictionaries discussed in Chapter 7.

Biographical information is sometimes required. Sources for today's chemists are indicated in Chapter 16, while Chapter 17 deals with biographies of famous chemists of the past. (This is about the only chemical field in which the popular, door-to-door peddled encyclopaedias are up-to-date!)

THE DETAILED SURVEY

In any survey or search the first requirement is a clear definition of the subject, nature and scope of the search. If the search is being carried out for someone else, it is essential to understand clearly the purpose of the survey. One of the major problems always is how far back to go and to what extent the nineteenth century literature can be ignored. Decisions on these problems will be tempered by the searcher's experience. It is possible to search back to 1871 with the aid of the precursors of *British Abstracts* or earlier with *Chemisches Zentralblatt* (*Chemical Abstracts*, of course, started in 1907). The pitfalls of multiple publication, anonymity, frequent changes of journal name, etc., which bedevil a search of the chemical literature published prior to 1875, have been discussed by Dyson (1961). In most cases an adequate summary of the nineteenth century literature can be extracted from either *Beilstein* or *Gmelin*.

Before 1930 the coverage of the literature by *Chemical Abstracts* was much less complete than that by *Chemisches Zentralblatt* and thus one should always search *both* for pre-1930 references.

If one can relate the object of the search to a particular compound, or even to several compounds, then one merely turns to *Beilstein* in the case of organic compounds for the work published up to 1930, and in some cases to 1960 (Chapter 11), or to *Gmelin* for inorganic ones

(Chapter 9). Up-to-date information can similarly be readily obtained via the appropriate Registry Numbers if one has access to the CAS data base.

Chemical Abstracts is the source of information published since the end of the period covered by either *Beilstein* or *Gmelin* until the current year. Because of the delays inherent in *Chemical Abstracts*, one must search through the previous years' issues of one of the alerting services described in Chapter 5 (preferably *Chemical Titles*) to bring the survey up to date, or perhaps get a computer search done through one of the on-line services described on pages 85–86.

A most useful feature of *Beilstein*, and one which is often not fully realized, is that the classification system brings a large number of closely related compounds together within a few pages of each other in the work. Thus, if one browses over a few pages on either side of the entry of interest, one can often get much useful information and ideas for further work. If one is likely to use *Beilstein* only occasionally, then perhaps it is unnecessary to learn the classification scheme and one can manage with the (1956) indexes for long-known compounds. There are, however, some substances which will be much quicker to locate by the system than by the indexes. A word of warning on using the indexes: trivial names are frequently used (e.g. 1,5-diaminopentane is listed as cadaverin).

Nomenclature also presents a problem in a rapidly expanding field such as antibiotics, where three or four groups of workers have each used a different name for the same substance. Thus one of the advantages of cumulative indexes is that a standardized nomenclature is used (see Chapter 5). Chapter 6 starts with a discussion of organic nomenclature – always a problem for the non-organic chemist.

Where the object of the search is a process, property, measurement technique, theoretical concept, etc., which is not easily identifiable with a particular compound, considerably more background knowledge is required than when searching for information about a compound. This should be acquired before going to the subject indexes of *Chemical Abstracts* so that the check list of headings under which one is going to search is as complete as possible. The technique of carrying out the search is described in Chapter 5 in the section on methods of using *Chemical Abstracts* and its indexes. Instructive examples of a search for a suitable analytical procedure and of an organic search have been given by Ridland (1960) and by Hancock (1968).

How does a chemist discover what relevant patent specifications have been published in his field of interest? The answer to this question is almost the same for patent literature as for any other kind of literature – he uses *Chemical Abstracts, Chemisches Zentralblatt*, etc. There are, however, a few reference sources peculiar to patent material. The most important of these for British readers are the *Classified Abridgements* or the Derwent services (if subscribed to) already mentioned in Chapter 14. In the former, subjects are broadly separated into a number of groups and within each group there is a finer sub-classification. The abridgements are arranged in volumes each of which has a detailed index of the numbers of each specification falling within a sub-class heading. For further instructions for dealing with certain patent information problems, see *Exercise 11* and the *Notes on the Solutions to the Practical Exercises.*

Most topics are covered by several possible abstracting services, as will be seen from the guides to such services described in Chapter 5. Tests on abstracting services by Martyn and Slater (1964) have shown that for a given subject there will be one service which covers it much more completely than the others. The coverage by the best service may, however, be much less than 100%. There seems to be no short-cut to discovering the best service for a particular subject other than carrying out the rather lengthy tests described by Martyn and Slater. Experience naturally helps one decide which service(s) to use, but fortunately for the chemist *Chemical Abstracts* is usually an adequate choice.

If one's search reveals a recent bibliography, before accepting it as a substitute for further searching, one should do a spot check on the references to the last two years' work covered by it. For obvious reasons, it is very difficult to produce a bibliography which is completely up to date at the time of publication (see Chapter 7).

When one's search has been completed within the limits one set at the start, a bibliography should be produced which should indicate the sources searched. A list of relevant review articles should be included so that anyone coming into the field can quickly assimilate the necessary background knowledge. These remarks apply whether the bibliography is to be published widely, included in a thesis or merely deposited in the firm's information file.

APPLIED CHEMISTRY

The most useful single-volume compilation for the production chemist is Perry's *Chemical Engineer's Handbook* (see Chapter 8). Two other useful quick reference works are *Riegel's Handbook of Industrial Chemistry* (J. A. Kent, Ed., 7th edn, Van Nostrand Reinhold, 1974) and Faith, Keyes and Clark's *Industrial Chemicals* (F.A. Lowenheim and M. Moran, 4th edn, Wiley, 1975), which reviews (US) processes and economics. Kirk and Othmer's or Ullmann's encyclopaedias (see Chapter 7) are invaluable reference works for the chemical engineer as well as the industrial chemist. There is, however, one work which is virtually a treatise in this field. This is *Chemical Engineering* (J.M. Coulson and J.F.R. Richardson, Eds, Pergamon Press, 1978—79). Some of the five volumes have a second or third edition and now use SI units. A chapter on other recent chemical engineering books will be found in *Use of Engineering Literature*.

Much of the primary publication on industrial chemistry is in the form of patents. The patent literature as a source of information is discussed at length in Chapter 14.

Much useful information is also to be found in such journals as *Industrial & Engineering Chemistry, Chemical Age* and so on. One should also peruse *Chemical & Engineering News, Chemical Week* and *European Chemical News*. This field is, of course, a commercial publisher's paradise. There are at least two or three 'technological glossies' for each industry and one should endeavour to scan those appropriate to the particular industry with which one is connected. Some articles in this type of publication are not abstracted by *Chemical Abstracts* but may be located through the *Engineering Index, Chemical Industry Notes* or, in the case of British papers, through the *British Technology Index*. A bibliographical guide, *Chemical and Process Engineering Unit Operations* (K. Bourton, Macdonald, 1967) covered the literature from 1950 to 1966. It was an annotated bibliography of journal articles (mainly reviews), books and reports noting the number of references for each item. A well-documented biennial review of progress is *Fortschritte der Verfahrenstechnik* (Verlag Chemie, 1952—   ). A number of consultants, especially the larger American ones, publish surveys of the scientific, technical and economic background to particular chemical processes for limited circulation among their clients. (Needless to say, such surveys are very expensive.) For example, Noyes Development Corporation have produced a series, *Chemical Process Reviews* (1967—69), which covers polymers, industrial gases, pesticides, organic halogenation and oxidation processes and products, synthetic organic nitrogen compounds, etc. Now called Noyes Data Corporation, they have published nearly 150 reviews, often based on US patents.

Government regulations for the production and use of chemicals (see Chapter 15) intimately concern the industrial chemist. Chapter 10 covers radiological hazards, and other aspects of health and safety are discussed in the next section.

Some elementary sources of information on the financial structure of industry are discussed in the final sections of Chapters 15 and 16.

HEALTH AND SAFETY

In the UK the enactment of the Health and Safety at Work, etc., Act 1974 has increased the obligations of employers to provide information for their employees and for the general public, in addition to obliging them to avoid risks to health and safety of these two groups. It appears to replace and extend the various Factories Acts, and brings together under the Health and Safety Executive inspectorates, research establishments and activities previously scattered in several Government Departments. The information requirements of the Act are discussed by Locke (1976). The Executive has established a large library of slides and films in London. It collects statistics, publishes reports and issues consultative documents. Maclaren publish the controlled circulation monthly *Health and Safety at Work* (1979–).

American regulations are summarized in the Commerce Clearing House's *Guidebook to Occupational Safety and Health* (1974).

The International Commission on Radiological Protection has produced several standards in this field. In 1966 it laid down maximum permissible exposure levels in its *Recommendations* (ICRP Publication 9). The International Atomic Energy Agency in Vienna produce a series of publications on health and safety, and codes of practice and maximum permissible dose rates have also been published by the UKAEA and are listed in the appropriate HMSO Sectional List. The Pergamon journal *Health Physics*, the organ of the Health Physics Society, publishes original papers in this area. An extensive review of ICRP publications and related material is L. S. Taylor's *Radiation Protection Standards* (CRC/Butterworths, 1971) (see also Chapter 10).

The standard work on safety aspects in the chemical and allied industries is *Dangerous Properties of Industrial Materials* (I. N. Sax, 4th edn, Reinhold, 1975). *ASTM Standards* are also widely used. Elsevier published a series of *Monographs on Toxic Agents* (E. Browning, Ed., 1960–). Other books are noted in *Health & Safety Literature: A Selective List of Material Held by the SRL* (J. Gaworska and D. King, BLSRL, 1977).

Unexpected hazardous reactions do occur in chemical laboratories and are normally scattered through the correspondence columns of several news or educational journals, especially *Chemistry in Britain* and *Journal of Chemical Education*. These reports have been very difficult to monitor and collate but L. Bretherick has made a detailed study and recently produced the *Handbook of Reactive Chemical Hazards* (2nd edn, Butterworths, 1979). Earlier N.V. Steere compiled the *Handbook of Laboratory Safety* (2nd edn, CRC Press, 1971) and published extracts from several US company safety manuals in *Journal of Chemical Education* from 1969 to 1970 under the heading 'Safety in the Chemical Laboratory', which is a regular feature in this journal.

The UK Chemical Industries Association has recently set up the Chemical Industry Safety and Health Council, which publishes a quarterly journal, *Chemical Safety Summary*, an irregular *Newsletter*, codes of practice and other monographs.

Much of the widely scattered material is noted in two abstracting services. A general service is *Safety Science Abstracts* (Cambridge Scientific Abstracts), covering 20 000 abstracts per year, including patents and reports. It covers instrumentation, education, prevention and legislation. Specifically for the chemist, *Chemical Hazards Abstracts* started as an experimental service from UKCIS in 1976 on microfiche. It is now one of the *CA Selects* titles (*see* page 68). UKCIS also provides several macroprofiles in this area ranging from Environmental Pollution to Recovery and Recycling of Waste.

PHYSICS, BIOLOGY AND OTHER PERIPHERAL SUBJECTS

The literature of physics is reasonably well documented, though not so well as that of chemistry. The most recent guide is *Use of Physics Literature* (H. Coblans, Ed., Butterworths, 1975). Parke's *Guide to the Literature of Mathematics and Physics* (2nd edn, Dover, 1958) performed a most valuable service in these fields and also covered some of the more mathematical works of engineering science but is now somewhat dated. *How to Find out in Mathematics* (J. E. Pemberton, 2nd edn, Pergamon Press, 1971) includes literature on computers and operations research and is generally considered the best of the Pergamon literature guides so far. A good up-to-date guide is *Use of Mathematical Literature* (A. R. Dorling, Ed., Butterworths, 1977). Also useful is J. A. Greenwood and H. O. Hartley's *Guide to Tables in Mathematical Statistics* (Princeton University Press, 1962). A very brief guide to statistical literature is included in Chapter 8 of *The Use of Biological Literature*.

There are two useful quick reference sources in physics; the shorter one is *Handbook of Physics* (E. U. Condon and H. Odishaw, Eds, 2nd edn, McGraw-Hill, 1967) and the larger is the Pergamon *Encyclopaedic Dictionary of Physics* (9 vols) (see also Chapter 8). The physical chemist engaged in building and maintaining electronic equipment will find the *Handbook of Electronic Circuits and Components* (1974) by J. D. Lenk, who has produced this and several related manuals for Prentice-Hall, an invaluable work to have in the laboratory. *How to Find out in Electrical Engineering* (J. Burkett and P. Plumb, 1967) is another Pergamon guide written by librarians rather than by practitioners in the subject. The only reasonably good and up-to-date guide covering this and other engineering fields is *Use of Engineering Literature* (K. Mildren, Ed., Butterworths, 1976). Those concerned with metals will find the *Metallurgical Dictionary* (J. G. Henderson and J. M. Bates, Reinhold, 1953) useful for quick reference and will also find much of the physical data they need in the *Metals Reference Book* (5th edn, C. J. Smithells, Ed., Butterworths, 1976).

The chemist is probably more concerned with the chemical rather than geological aspects of minerals and will therefore find Hey's *Index of Mineral Species and Varieties* (British Museum, 2nd edn, 1955, Appendix, 1963) still of interest. This index is arranged chemically and has an alphabetical list of accepted mineral names and synonyms. Processing, production figures, etc., of about 100 minerals are dealt with at length in *Minerals for the Chemical Industry* (S. J. and M. G. Johnstone, 2nd edn, Chapman and Hall, 1961). The best guide to the geological literature is *Use of Earth Sciences Literature* (D. N. Wood, Ed., Butterworths, 1973). *A Guide to Information Sources in Mining, Minerals and Geosciences* (A. Kaplan, Interscience, 1965) lists more than 1000 organizations and 600 periodicals, directories, etc. It is, however, more an address list than a literature guide.

To judge from the current one-third biochemical content of *Chemical Abstracts*, a knowledge of the biological sciences is as important to the chemist as a knowledge of the physical sciences. A useful quick reference work is *Encyclopedia of the Biological Sciences* (P. Gray, Ed., Reinhold, 1961). This and other quick reference sources are discussed in Chapter 8 of *The Use of Biological Literature*. This is still the only guide which covers the whole field of biological literature. A manual widely used by American zoology students is *Guide to the Literature of the Zoological Sciences* (R. C. Smith and R. H. Painter, 7th edn, Burgess, 1967). The *Dictionary of Microbiology* (M. B. Jacobs *et al.,* Van Nostrand, 1957) helps one to overcome the vocabulary barrier, as, of course, does the shorter and more general Penguin *Dictionary of Biology* (M. Abercrombie *et al.,* 6th edn, 1969).

The chemist working in a medical field has more literature problems than most, that of terminology being one of the lesser ones. Perhaps the chemist will find L. T. Morton's *How to Use a Medical Library* (6th edn, Heinemann, 1979) helpful, as it lists medical periodicals, abstracting services and bibliographies, but far more detail is given in *Use of Medical Literature* (L. T. Morton, Ed., 2nd edn, Butterworths, 1977). *Butterworth's Medical Dictionary* (M. Critchley, Ed., 2nd edn, 1978) has been very favourably reviewed and provides an authoritative solution to the terminology problem. *The Chemist's Dictionary of Medical Terms* (Anon., 8th edn, Morgan-Grampian, 1967) is intended for use in pharmacies but will also prove useful to research chemists. Much fuller information is given in *Cyclopaedic Medical Dictionary* (C. W. Tabler, Ed., 12th edn, Davis, 1973).

Outlined above are a selection of the major literature guides and reference works for some of the areas of knowledge where the chemist may occasionally find himself. The time-honoured expedient for getting information is, of course, to ask an expert in the relevant field. This section has been written not so much as a substitute for this procedure as for the many occasions when a good library is more accessible than the expert. For subjects not covered in this section the reader is referred to A. J. Walford's *Guide to Reference Materials* (Vol. 1, *Science & Technology*, 3rd edn, Library Association, 1973), which contains about 3000 annotated entries classified by the UDC, or the 10 000 entry *Guide to Reference Books* (E. P. Sheehy, Comp., 9th edn, American Library Assn, 1975), formerly known as *Winchell's Guide*. The reader who detects eccentricities in these two guides should remember that they are compiled by librarians rather than by users, thus reinforcing the plea for book lists produced by subject experts which is made in Chapter 7. A useful general introductory guide written by scientists is C. C. Parker and R. V. Turley's *Information Sources in Science and Technology* (Butterworths, 1975).

COMMUNICATION TO OTHERS

Much technical reading is undertaken so that a digest of the information obtained may be passed on to others, less well qualified technically than the reader, be they undergraduates or commercial management. For the former group, one must digest portions of suitable monographs and review articles (which are discussed in Chapter 7); for the latter, one must be familiar with the literature mentioned in the applied chemistry section of this chapter and especially the appropriate trade statistics, financial situation, and so on. (Daily reading of the *Financial Times* is most important in this context; one should also try to read regularly the periodicals mentioned in the last section of Chapter 15.)

For those concerned with industrial training schemes the film may well be preferred to the formal lecture for instruction. Suitable films may be located through the *RIC Index of Chemistry Films* (6th edn, 1970), which lists 1600 films and 400 film strips. New additions are noted in *Education in Chemistry*. Those particularly interested in this medium will find the UNESCO periodical *Scientific Film* a most useful source of current information. Also of interest is *Films in Higher Education and Research* (P. D. Groves, Ed., Pergamon Press, 1966), the proceedings of a 1964 conference.

It has often been said that there is little point in a scientist finding out anything, whether it is at the bench or in the library, if he cannot communicate his findings to other people. Several books have been written to help the poor, unlettered technical man express his ideas adequately. Even some well-known chemists have gone into print in this field. Technical writing is, however, little different from any other form of English composition. The same Golden Rule applies: It is imperative that a predilection for the polysyllabic be vouchsafed minimal transudation, i.e. don't use two long words when one short one will do. The finest book on English usage (and one of the cheapest) is Sir E. Gower's *The Complete Plain Words* (2nd edn, revised by Sir B. Fraser, HMSO, 1973), which is a reconstruction of Gower's famous *Plain Words* and *ABC of Plain Words*. If one, however, is stuck for the right word, Roget's *Thesaurus* (now available in a Penguin edition) will probably be more useful than a normal dictionary. In it words are arranged according to the ideas which they express, as well as alphabetically. The correct medical term, however, is found more readily from *Reversicon, A Medical Word Finder* (J.E. Schmidt, Thomas, 1958).

The chemist preparing a technical paper for the first time will find the *Handbook for Chemical Society Authors* (1960) very helpful. It contains IUPAC rules for inorganic and organic nomenclature, notes on English usage and marking copy for printing, and so on. Unfortunately, the Chemical Society's method of citing references and abbreviating journal titles is little used outside its own publications. Many journals and publishers issue notes for guidance of prospective authors.

Those who have to present papers at a meeting will find much useful information on preparation of slides, speaking and many other aspects of the presentation of their material in *Oral Communication of Technical Information* (R. S. Casey, Reinhold, 1958) or more briefly in an article by Williams (1965). Sometimes, unfortunately, the actual delivery of a paper at a conference receives all too little attention from the author. The Golden Rules of oral communication are, however, neatly phrased in the old saw: Stand up, Speak up and Shut up!

KEEPING UP TO DATE

This is perhaps the most difficult literature problem of all, unless one is pioneering in a field in which very few others are working. The procedure adopted for keeping abreast of the literature depends, of course, on the individual's circumstances and interests. A few generalized pointers can, however, be given. Most of the steps to be taken will by now be obvious to anyone who has read his way through this book.

Whatever one's field of interest may be, *regular* perusal of the appropriate journals, followed by systematic and immediate recording of any relevant references directly onto one's reference cards is the only hope of keeping up to date. The appropriate journals fall into four classes, the 'quick publication' medium, the main or 'prestige' journals of one's sphere of interest, the alerting service and, finally, comprehensive abstracts publications. A method of selecting the most rewarding primary journals using the *Science Citation Index* is suggested on pages 80–81.

'Quick publication' journals are most important for study, as it is in these that preliminary communications of work, which may not be published in full for many months or even years, appear. At one time the most important of these was *Nature* but publication of the 'Communications to the Editor' took longer and longer until around 1960 they were delayed as long as full-length articles in a major journal. This has now been rectified and 'Letters to the Editor' are once more a useful source of rapid preliminary communication, especially on areas which are peripheral to chemistry. The current delay is about two months from acceptance. Few of the 'quickies' in the Reports section of *Science* are of direct chemical interest. In the past two decades *Chemistry & Industry* has become a most useful, international source of preliminary communications in all fields of chemistry. Publication is currently three to six months after receipt of the manuscript, which is, of course, refereed. Another important English-language publication in this class is *Journal of the Chemical Society Chemical Communications* (three to four months' publication delay). Two important German journals must be noted which contain preliminary communications. They are *Naturwissenschaften* (about three months' delay) and *Angewandte Chemie* (one to three months' delay for the original German edition, with an additional one to two months' delay in the international edition). There is a considerable demand for space in such publications and in recent years a number of commercial publishers have brought out such journals. In order to minimize publication delays they are often photolithographed from the typescript once it has been accepted by the referees. Usually no proofs are issued. Using such methods *Chemical Physics Letters* (1967–) claims to publish communications within 14 days of acceptance. One of the oldest of these journals is *Tetrahedron Letters* (organic chemistry); *Analytical Letters, Polymer Letters*, etc., have already been mentioned in previous chapters. Preliminary communications are also sometimes made in conference papers (see Chapter 7). There is, however, no doubt that, despite the efforts of the editors of the more responsible journals, this medium is abused by publication-hungry scientists whose future promotion and grants depend on the quantity of their publications, since this is more easily assessed than quality.

The main journals are discussed in Chapter 3 and one should, depending on the facilities available, try to peruse the title pages of at least three or four which are appropriate to one's interests as well as the journals of the British and American chemical societies.

The list of headings under which one searched indexes in one's original literature survey should be used as a basis for making a list of key words under which one searches *Chemical Titles* each fortnight. Each month one should also make a point of reading 'Highlights' in *Chemistry & Industry* (q.v.).

Finally, the appropriate sections in *Chemical Abstracts* should be scanned each fortnight for the papers which slipped through the net formed by the previous three procedures.

One should not keep too strictly to one's subject when looking through journals but permit oneself to browse a little. Browsing is much underestimated as a source of inspiration for further work.

There is a fifth category of periodical which one should browse through whenever time permits. This is the news journals such as *Chemical & Engineering News* and *Chemistry & Industry*. If possible, one should try to keep abreast of the purely commercial side of chemistry through *Chemical Industry Notes*, etc., or the type of publication discussed in the final sections of Chapters 15 and 16. The news journal is rightly of prime importance to the technologist but it is hoped that no professionally active chemist would confine his reading (if any) to this category as did a significant number of technologists in the electrical and electronics industries (Calder, 1959). It is hoped that by discussing the many different types of chemical literature, this book will make the chemist more literature-conscious and enable him to use it efficiently and effectively.

REFERENCES

Calder, N. (1959). *What They Read and Why*, DSIR
Dyson, G. M. (1961). *Advances in Chemistry Series, No. 30*, pp. 83–91
Hancock, J. E. H. (1968). *J. Chem. Educ.,* **45**, 336
Locke, J. H. (1976). *Aslib Proc.,* **28**, 8
Martyn, J. and Slater, E. (1964). *J. Docum.,* **20**, 212 (see also Martyn, J. *ibid.,* **23**, 45, 1967)
Ridland, W. (1960). In J. L. Ward, *Library Services for Chemists*, Royal Melbourne Technical College Press, pp. 38–49
Williams, P. C. (1965). *Inst. Biol. J.,* **12**, 65

Appendix 1

Suggestions for practical work

A set of exercises covering most types of problems encountered when using the chemical literature is appended. They have all been used on recent postgraduate courses run at Strathclyde, Syracuse and City Universities and are designed to give the student experience in handling the literature so that when he has a real literature problem of his own, he will recognize its existence and be able to solve it.

Instructors should allocate only one or two questions of each type in an exercise to each student. They should, of course, first check that the requisite books, etc., are available, as it is extremely demoralizing for the student if it is impossible for him to do the exercise set. It is most important that those interested in organic chemistry attempt the physical and inorganic chemistry problems and vice versa, as no-one knows when a particular project will demand knowledge from a different branch of chemistry. Some notes on the solution of the exercises will be found at the end.

Questions marked * may prove rather difficult for non-chemists.

Exercise 1

Periodicals

(1) Give the full name, country of origin, year of first issue, language, frequency of appearance, volume number(s) for 1961, and name the nearest library which takes:

(a) *Annalen*
(b) *Bulletin of the French Chemical Society*
(c) *Arch. Pharm.*
(d) *Org. Syn.*

(e) *Ber.*
(f) *Analyt. Chim. Acta*
(g) *J. Colloid Sci.*
(h) *Rev. Agr. Agron.*

(2) Give:
   (a) Date, and place of the 5th Conference of the International Union of Pure and Applied Chemistry.
   (b) Date of volume 79 of *J. Am. Chem. Soc.*; volume 17 of *J. Prakt. Chem.*; cumulated indexes to *J. Chem. Educ.*
   (c) Address of *Oceanus*; *Blech*; *DIA Boletin Tecnico.*
   (d) Two sets of abbreviations for (i) *Journal of the Electrochemical Society, New York*, (ii) *Chemical Society of Japan, Bulletin.*

(3) Give full references for:
   (a) A paper published in 1953 on the bioassay of relaxin.
   (b) A paper by Everson in 1956 on polyhydric alcohols.
   (c) Initials of Young, who wrote a paper on inspection of plant and equipment for the chemical industry in 1960. What is his address?
   (d) W. F. Harrington and Schackman, *J. Amer. Chem. Soc.*, **75**.
   (e) Lettre and Fernholz, *Journal of Physiological Chemistry*, **278**, 175–200 (1943).

Exercise 2

'Chemical Abstracts' and its indexes

(A1) Using the 1947–56 decennial formula index, find a paper on the effect of deuterium ethyl sulphonate on aliphatic hydrocarbons.
(A2) Find an article, between 1937 and 1946, on the reaction of morpholine with

$$
\begin{array}{c}
CONH \\
ClCH_2CH_2CH \diagup \diagdown CHCH_2CH_2Cl \\
NHCO
\end{array}
$$

(A3) What hydrated salts of $p\text{-}IC_6H_4SO_3H$ are recorded as being prepared in *Chemical Abstracts* for 1927–36? Give the reference of the papers.
(A4) What compound can be obtained by the reaction of $CH_3COCH_2CH_2OCCH_3$ with NAN_3? Where is it described since 1940?

(B1) Complete the following reference:
'Effect of .... on broiler-starter rations'. W. K. Warden and P. J. Schaible. .... *Sci.*, 1958, 37,......................

(B2) Complete the following reference:
'The influence of the degree of .... on the characteristics of .... and a new method for determining its .... content'. L. I. ...., T. A. Kiparisova and A. M. Komraz. .... 1957, No. 6.......................... .

(B3) Complete the following reference:
'Dielectric absorption of ....'. .... and A. H. Price, 1955, ... 2204.

(B4) Complete the following reference:
'Method of Nicloux for determining ....'. G. F. De Gaetani, *Boll. Soc. Ital. Biol. Sper.*, 1940, ...

(B5) R. O. Cinneide published a paper in *Nature*, 1955, page 47, for which experimental details were to be published elsewhere. Has this been done and if so, what is the reference?

(C1) 'The structure of films of long chain materials on copper'. Compile a complete list of references from the published literature up to 1960.

(C2) Compile a list of references, including titles of articles in the published literature before 1950 on: 'Use of glycerine as a lubricant'.

(C3) What articles were published before 1950 on: 'Effect of smoking tobacco on blood pressure'?

(C4) What articles have been published on: 'Effect of lithium 12-hydroxystearate as a thickener'?

Series A questions are based on *Chemical Abstracts* formula index.
Series B questions are based on *Chemical Abstracts* author indexes.
Series C questions are based on *Chemical Abstracts* subject index.
(Questions A1 to B5 are shorter than the others and should be attempted first.)

Exercise 3

Reviews and monograph series

(1) List review periodicals or periodicals containing reviews which cover the following fields:

(a) Biochemistry	(e) Analytical chemistry
(b) Protein chemistry	(f) Metallurgy
(c) Physical chemistry	(g) All fields of chemistry (two
(d) Applied chemistry	examples)

(2) In 1955 a review entitled 'Inclusion compounds' was published. Give: Reference; number of references cited in bibliography; abstract reference, if any.

(3) Suggest a review on 'Liesegang phenomenon'. Where was it first described in a journal?

(4) Suggest a review on 'the cleavage of the carbon-sulphur bond in sulphur compounds'. Does it mention a book on the organic chemistry of sulphur by Suter? If so, what is its reference?

(5) (a) What is tiglic acid?
    (b) Where can one find a review on it?
    (c) What is its relation to angelic acid?

Exercise 4

Miscellaneous

(1) At the third International Bread Congress, Professor K. Hess delivered a lecture on 'The endosperm proteins of wheat'.
    (a) What was the date and venue of the Congress?
    (b) Give an address through which Professor Hess may be contacted.
    (c) Give reference to a *review* article on the above subject.

(2) Is the following article, Tartarini, *Ann. Chim. Applicata*, 1933, **23**, 367–72, listed in the Author Index of *Chemical Abstracts*? If yes, give volume number, page and year of author index.
Is this article listed in the author index of *Chem. Zbl.*?
If yes, give vol. number, page and year of author index.
Is this article listed in the author index of *British Abstracts*?
If yes, give year and page of author index.
Give name of author(s), title (in English or German) and correct citation.

(3) An article on saccharin in Thorpe's *Dictionary of Applied Chemistry* (4th edn) incorrectly cites a paper on the inactivation of the zinc component of insulin by various sulphones. What is the correct citation?

(4) Select at random three books from the book lists appended to Chapter 17 and give publisher, date and place of publication.

(5) Make a list of papers referring to light scattering in pure water (often Brillouin or Rayleigh scattering) published since the beginning of 1966.

One of the authors interested in light scattering in water is M. G. Cohen. Has he published any other papers on light scattering in the period 1966 onwards?

(You can use this question to compare the coverage of *Physics Abstracts* and *Chemical Abstracts* if you have time.) Locate the key references by Brillouin and by Rayleigh and see whether any additional references are located from the *Science Citation Index*.

(6) Have any conferences on Mössbauer spectroscopy taken place in the last three years?

(7) Recently you read a review on micelle formation presented at an international conference in 1960. Now you can remember that the paper had 148 references, but you have forgotten the author and the title of the conference. Try and find these again, noting where you find them. (Papers from conferences usually take a year to reach secondary sources.)

Exercise 5

Compilation of a bibliography

(1)† List as many references as you can find published since 1972 which deal with ....
Under which section in *Chemical Abstracts* does the majority of these articles appear?
State in which journals you have consulted the annual index when searching for the latest references.
Have you searched *Chemical Titles*? If so, under which keywords?
What 'target' papers would you use in a search in *Science Citation Index*?
List any other steps taken?

Exercise 6

Physical chemistry and physicochemical data

(1)†† Give the ICT (*International Critical Tables of Numerical Data*) value for the melting point of ....
Give the value, with reference, for any later determination of this melting point.

(2)§ List two references in which laboratory directions are given for making the following physical chemical measurement: ....

(3)§§ Where can the vapour pressure data on .... be found in:

---

† *Suitable topics for Q. 1:* Isocyanate toxicity, organic halogen exchange reactions, thermochromism, copper pyridyl alkanoates, thermodynamics of adsorption, photolysis of organometallic compounds, heterometric determination of Ca, biosynthesis of rubber, electroplating of silver.

†† *Suitable substances for Q. 1:* Beeswax, crocetin, litharge, indene, willemite, wulfenite, UI_4, ephedrine, BeO, ceresin, ZrO_2, germanium.

§ *Suitable measurements for Q. 2:* Optical rotatory dispersion, intrinsic viscosity of polystyrene, heat of solution of KBr, specific heat at constant pressure of propane, retractive index of a dilute protein solution.

§§ *Suitable substances for Q. 3:* Propane, DCN, UCl_4, quinoline, mercury dimethyl, perfluorobutane, phosgene, sodium.

(a) *ICT:*

(b) *Landolt–Börnstein:*

(c) *Tables Annuelles de Constantes et Données Numeriques de Chimie, de Physique, de Biologie et de Technologie:*
(If not listed, write 'not reported.')

Exercise 7

Inorganic chemistry and radiochemistry

(1) Obtain information on the nuclear data (half-life, type and energy of disintegration) and recommend the most appropriate radioisotope for a research investigation (examine the cost and availability if possible) involving C, Ca, Sr, I.

*(2) Recommend a method for the determination of radiostrontium and radiocalcium in drinking water supplies.

(3) Obtain the latest information on the following topics:
  (a) production of nitric acid by ionizing radiation;
  (b) preparation of a CoO catalyst supported on Al_2O_3;
  (c) preparation and properties of elementary Cm, Np;
  (d) preparation and properties of $Tc_2(CO)_{10}$;
  (e) forms and structures of Fe_2O_3;
  (f) zirconium phosphate as an ion-exchanger;
  (g) valency states of Re;
  (h) the chemical toxicity of Be;
  (i) phosphine complexes of metals;
  (j) TBP complexes of cations.

(4) A summary of the chemistry of lithium may be found in the following works:
  (a) Mellor: *Treatise on Inorganic and Theoretical Chemistry*, vol.   pp.
  (b) Gmelin: *Handbuch der anorganischen Chemie*, 8th edn, Vol.   pp.
  (c) Pascal: *Traité de chimie minérale*, Vol.   pp.

(5) What is the chemical name and the formula of the mineral gadolinite?

Exercise 8

The use of Beilstein's 'Handbuch'

Locate and record the positions of entries in all volumes (Hauptwerk and supplements) for the following: (1) 1,5-diaminopentane, (2) 2-methylquinoline, (3) 1-hydroxynaphthalene-4-sulphonic acid, (4) 2-bromoethanol, (5) phosgene, (6) phenol-formaldehyde resins,*

(7) saccharin ($C_7H_5NO_3S$), (8) ethylene glycol dibenzoate, (9) octadecane 1,12-diol distearate, (10) $CH_2ClCF_2OCH_2CH_3$,* (11) $CH_3CH(N_3)$ $COOCH_3$, (12) *N*-benzoylbenzylamine, (13) sucrose, (14) succinic acid, (15) penta-*O*-acetyl-α-D-fructopyranose, (16) CD_3CF_3.

Use any location procedure you wish. In certain cases indexes will be useless, in others they will give results quickly. It is good experience to locate some of the above independently by both methods (indexes *and* classification system) and to time both, so determining for yourself the relative merits of each.

Where you do locate by use of the system, record your classification breakdown (Div., Subdiv., Class, etc.) along with the location.

Exercise 9

Organic chemistry

*(1) Give the sources from which the following information can be obtained:

(a) melting point;
(b) melting point of hydrochloride;
(c) chemical properties;
(d) uses in synthesis of:

*(2) Obtain details of the synthesis, chemical and physical properties of $NH_2CONHNO_2$.

(3) Locate a reference source which gives in some detail the solvent properties of $(CH_3)_3CCH_2CH_3$.

*(4) Locate a review article on the Gattermann reaction for the synthesis of aldehydes, and from it obtain details of an original paper referring to the preparation of

Find the melting point of this compound, and those of its acetyl derivative and oxime.

(5) Find the systematic name for the dyestuff violanthrone.

(6) The 'Literature Preparation' is a time honoured method of introducing students to the chemical literature in Central European Universities. Students are required to consult the original literature for practical instructions for preparations such as:
   (a) *N*-bromethyl phthalimide;
   (b) 2,4,6-trichlorphenoxy acetic acid;
   (c) haem from ox blood;
   (d) 2-butoxytetrahydropyrane from tetrahydrofurfuryl alcohol;
   (e) 2,6-dinitro-4-methoxy phenol from hydroquinone;
   (f) fluorenone from phenanthrene.

See further *Organikum, Organisch-chemisches Grundpraktikum* (Author-collective from Dresden Technical University, 7th edn, VEB Deutscher Verlag der Wissenschaft, East Berlin, 1967).

Exercise 10

Analytical chemistry

(1) (a) What section of *Chemical Abstracts* is devoted to analytical chemistry?
   (b) What section of *British Abstracts* was devoted to analytical chemistry?
   (c) Select a 1951 article in *Analytical Chemistry* and compare the abstracts of this article in the two abstract journals listed above. (Describe type of abstract, promptness of appearance, clearness of expression, whether too short, not complete enough, etc.)
   (d) Do the same for an original article in *The Analyst*.
(2) List (giving full name, abbreviation, inclusive dates) one analytical chemistry journal published in each of the following countries: England; USA; Germany; Russia; Netherlands.
(3) Where was the International Symposium on Microchemistry (1958) held? Are the *Proceedings* published? If so give publisher and date of publication.
*(4) Recommend a simple method for the colorimetric estimation of sulphate in deep well water.

Exercise 11

Patents

(1) (a) About 1952 there was published a British Patent to Distillers Co. Ltd on the use of nickel catalysts in hydrogenation. What was the patent's number and did it cover reactivation of the catalyst above 200°C?

(b) About 1952 there was published a British Patent to Marchon Products Ltd on detergent compositions containing hydrophilic colloids such as methyl cellulose. What was the number of the patent and does a mixture of methyl cellulose, palmityl mono-ethanolamide and sodium lauryl sulphate fall within the claim?

(c) About 1957 there was published a British Patent to Dow Chemical Co. on stabilization of p-bromophenols. What was its number and did it cover the use of lead pyrophosphate as stabilizer?

(d) About 1957 there was published a British Patent to Nopco Chemical Co. on making vitamin-A-containing products. What was its number and does it cover heating whale liver oil with 0.1% hydroquinone to 250°C for 1 min and then rapidly cooling to room temperature?

(2) Are there British patents equivalent to USP 2 865 861; 2 865 958; 2 867 504, 2 870 019; 2 871 124; 2 894 912; 2 947 701?

(3)† (a) Make a list of the numbers of any British Patents on the catalytic hydrogenation of animal or vegetable oils and fats published in 1951–52.

(b) Make a list of the numbers of any British Patents on the production and purification of glycerol published in 1951.

(c) Make a list of the numbers of any British Patents on the production and purification of glycerol published in 1952.

(d) Make a list of the numbers of any British Patents on the production and purification of glycerol published in 1953–54.

(e) Make a list of the numbers of any British Patents on the production and purification of phenol or cresols published in 1951.

(f) Make a list of the numbers of any British Patents on the production and purification of phenol or cresols published in 1952.

(4) Assuming that you are maintaining an up-to-date watch for certain classes of Patents and wish to have them on the morning after date

---

† In Q. 3 an artificial restriction has been placed on the number of years to be searched in order to make the exercises completable in a reasonable time.

of publication, peruse the latest copy of the *Official Journal* (*Patents*) for patents concerning:

(a) the production, preparation and use of polyethylene and polypropylene;

(b) the production and treatment of steroids;

(c) organosiloxanes;

(d) nuclear reactors and fuels.

Record the name(s) of the Patentees (if any), and give date of filing of provisional application and of complete specification (where applicable). Note the main Index at Acceptance quoted, and check the index of Classification Numbers further on in the *Official Journal* to find whether any have been missed. Finally, turn up the actual specifications to see whether your original selection by title has given you the information you require.

Exercise 12

Quick reference queries

The following are actual queries received by an information unit. Answer and give reference to source of data.

(1) Toxicity and flashpoint of tetrahydrofuran.

(2) Flashpoint of styrene.

(3) Thermal conductivity of hydrogen fluoride gas.

(4) Specific heat of borax.

(5) Solubility of estrone.

(6) Formula of Freon 22.

(7) Solubility of methane in water.

(8) Density of liquid helium II.

(9) Percentage moisture content of raisins.

(10) Specific gravity of gin and rum.

(11) Iodine value of herring oil and herring meal.

(12) Specific gravity of sherry.

(13) Percentage composition of dried egg yolk.

(14) Specific heat of kerosene.

(15) Viscosity of boiling sugar.

(16) Calcium content of ox blood.

(17) Determination of naphthalene in solids.

(18) Sulphate limit of commercial nickel carbonate.

(19) Vapour pressure of benzene at $0°C$.

(20) Properties and characteristics of trichloroethylene.

(21) Details of furazolidone and its manufacture.

(22) Flashpoint of carbon disulphide.

(23) Information on uses of diaminodiphenylsulphone.

(24) Information on thyroxine.

Exercise 13

Assessing the relevance of journals of marginal interest

A scientist claims that several journals not taken by your library are relevant to his field. He has supplied at your request a list of ten key papers which define his (rather narrow) field of interest. Check the ten papers in the most recent *Science Citation Index* volume and note the names of the journals which cite any of these papers. Rank them in order of decreasing frequency of citation, with the number of citations. Check whether the requested journals are scanned by the *Science Citation Index* and if so note their rank in your list (if in fact they appear).

P. A. Anderson, *Phys. Rev.*, 1959, **115**, 533.

N. Cabrera and N. F. Mott, *Repts Prog. Phys.*, 1948, **12**, 163.

R. V. Culver and F. C. Tompkins, *Adv. Catalysis*, 1959, **11**, 67.

G. Ehrlich, *ibid.*, 1963, **14**, 255.

T. B. Grimley and B. M. W. Trapnell, *Proc. Roy. Soc. A*, 1956, **234**, 405.

P. M. Gundry and F. C. Tompkins, *Quart. Revs.*, 1960, **14**, 257.

N. B. Hannay, *Semiconductors*, Reinhold, 1959.

J. C. P. Mignolet, *Disc. Faraday Soc.*, 1950, **8**, 105.

Idem, *ibid.*, 1950, **8**, 326.

H. H. Uhlig, *Acta Met.*, 1956, **4**, 541.

Suggested marginal journals are *Czech. J. Physics* and *J. Electrochem. Soc.*

Exercise 14

Comparison of manual and KWIC indexing

Select at random six abstracts from *Chemical Abstracts* Vol. 64 (1966) Subject Index. Note indexing term and abstract reference. Look up each abstract reference and from the title of the paper decide what the KWIC index entries would be. (NOTE: Use the stop list of non-significant words in *Chemical Titles* to decide which words are not KWIC indexed.)

How many of the six subject entries are (a) identical with (including syntactical variations), (b) synonyms or paraphrases of, (c) no connection with, a KWIC index entry?

Exercise 15

Wiswesser Line-Formula Notation

Code into or decode from WLN the following:

(1) $CH_3COOCOCH_3$.

(2) $CH_3CH_2OCONHCH_2COOCH_2CH_3$.
(3) 4UU3.
(4) Z2VQ.
(5) Z1VO2S1.

(6)

(7)

(8) 3R.
(9) RV1R.
(10) ZMR CG.
(11) Zr BMR CQ.
(12) T6NJ.

Appendix 2

Notes on the solutions to the practical exercises

In most cases there will be more than one way of getting the required answer. General methods for doing this are indicated in the appropriate chapters and thus these notes serve only to direct the student to the appropriate reference tool and to point out any irregularities.

Exercise 1

(1) Consult *CASSI* or *List of Periodicals Abstracted by Chemical Abstracts*. The 1961 edition gives the volume number for 1961. The nearest (British) library to take the journal will be found from the *World List* (Chapter 3).
   Notes: (b) (incorrectly) translated title; (e) colloquial abbreviation; (h) Vol. 28 in 1961, journal is not in 1961 *List*, therefore one must search *Supplements* (bound with Author Index); it is in the 1962 *Supplement* or *CASSI*.
(2) (a) See *World List*.
   (b) A synchronistic table (see Chapter 3) must be used for the first two (unless volumes are readily available on the shelves).
   (c) See *CASSI (1907–74)*.
   (d) e.g. from *World List* and *CASSI*.
(3) (a) See *Biological Abstracts* 1954 Subject Index.
   (b), (d) and (e) Use Author Indexes of *Chemical Abstracts* ((d) see Q. 2 (b); (e) is an example of a translated journal title).
   (c) An obvious case for using *Engineering Index*! Check address from his later papers (or cf. Chapter 16).

Exercise 2

(A1) 'The action of sulphuric acid, ethane sulphonic acid and chloro-sulphonic acid on aliphatic hydrocarbons'. G. S. Gordon III and R. L. Burwell, *J. Amer. Chem. Soc.*, 1949, **71**, 2355–9; (*Chem. Abstr.*, **44**, 4409e).

(A2) 'Synthetic amino acids. Some reactions of 3,6-bis (β-chloroethyl)-2,5-diketopiperazine'. H. R. Snyder and M. E. Chiddix, *J. Amer. Chem. Soc.*, 1944, **66**, 1000–2; (*Chem. Abstr.*, **38**, 3984^3).

(A3) If $p\text{-IC}_6\text{H}_4\text{SO}_3$ equals X, hydrated salts are $\text{X}_2\text{Mg6H}_2\text{O}$; X_2Zn $6\text{H}_2\text{O}$; $\text{X}_2\text{Cd6H}_2\text{O}$; $\text{X}_2\text{SrH}_2\text{O}$. 'Aromatic sulphonates of elements of the 2nd group of the periodic system'. V. Cupr and J. Sirucik, *J. Prakt. Chem.*, 1934, **139**, 245–53; (*Chem. Abstr.*, **28**, 3325^1).

(A4) $(\text{CH}_2\text{CONHMe})_2$. 'Reactions with hydrazoic acid in sulphuric acid solution. X: The behaviour of γ-diketones'. G. Caronne, *Gazz. Chim. Ital.*, 1950, **80**, 675; (*Chem. Abstr.*, **45**, 8980e). Note: This may be more easily found by using the name 2,5-hexanedione in the subject index.

(B1) 'Effect of gibberilic acid on broiler-starter rations'. W. K. Warden and P. J. Schaible, *Poultry Sci.*, 1958, **37**, 490–1.

(B2) 'The influence of the degree of fermentation on the characteristics of kvass and a new method for determining its alcoholic content'. L. I. Chekan, T. A. Kiparisova and A. M. Komraz, *Trudy Vsesoyuz Nauch-Issledovatel Inst. Pivovarennoy Prom.*, 1957, No. 6, 114–21.

(B3) 'Dielectric absorption of triethylamine solutions'. Mansel Davies and A. H. Price, *J. Chem. Phys.*, 1955, **23**, 2204.

(B4) 'Method of Nicloux for determining ethanol in blood'. G. F. De Gaetani, *Boll. Soc. Ital. Biol. Sper.*, 1940, **15**, 750. Note: Abstract delayed by World War II.

(B5) No.

(C1) Subject index entries are under 'Copper – films on'.

(C2) Subject index entries are under 'Glycerol'.

(C3) Subject index entries are under 'Tobacco–smoke–effect on blood pressure'.

(C4) Subject index entries are under 'Octadecanoic acid–12-hydroxy–lithium salt'.

Exercise 3

(1) See BLLD's *KWIC Index . . . to Review Publications* (Chapter 7) or *Ulrich*.

(2) See *Chem. Abstr.* Subject Index, 1956.

(3–5) Locate the review (through the Cumulative Index to *Chemical*

Reviews, 1960) from which the subsidiary questions can be answered.

Exercise 4

(1) (a) Hamburg, 1955 (see *Chem. Abstr.* Author Index, 1958).
   (b) From *Chem. Abstr.* or *Addressbuch der deutschen Chemiker.*
       N.B.: Do not confuse him with Kurt Hess of Glanzstoff-
       Courtaulds, Cologne.
   (c) See *SCI Monograph No. 6*, 1959 or *Advances in Protein
       Chemistry*, 1945.
(2 and 5) Solutions are self-evident.
(3) Use *Chem. Abstr.* Author Index but check all authors if first
    author is not listed.
(4) Reference to a published library catalogue or to the *Cumulative
    Book Index* or appropriate National Bibliography should produce
    the required details. (See Chapter 2.)
(6 and 7) Check *Chem. Abstr.* Subject Indexes or guides mentioned in
    Chapter 7.

Exercise 5

(1) See Chapter 18 under *The Detailed Survey.*

Exercise 6

(1) Check for later determination in NBS *C500* etc. (or in *Chem. Abstr.*
    Subject Indexes).
(2 and 3) Solutions are self-evident.

Exercise 7

(1) See Chapter 10 under *Radioisotopes.*
(2) Search *Nuclear Science Abstracts* or *Guide to UKAEA Documents*,
    etc.
(3) (a), (c), (d), (h) and (j) Search *Nuclear Science Abstracts.*
    (b), (f) and (i) Search *Chem. Abstr.* Subject Indexes.
    ((f) See also *Journal of Inorganic and Nuclear Chemistry*, and (i)
    see also *Ann. Reports*.)
    (e) and (g) See *Gmelin*, etc.
(4) NOTE: Supplement to Mellor.
(5) Check in any of treatises of Q. 4 or in Hey's *Index of Mineral
    Species* (see Chapter 18 under 'Peripheral subjects').

Exercise 8

Entries	Comments
(1) 4, 266; I, 421; II, 708; III, 588.	Indexed under cadaverin
(2) 20, 387; I, 148; II, 238	
(3) 11, 271; I, 64; II, 154; III, 541*	*Derivatives only
(4) 1, 338; I, 170; II, 337; III, 1359; IV, 1385.	
(5) 3, 13; I, 7; II, 12; III, 31; IV, 31.	
(6) Various, inc. 6, 124; I, 75; II, 131: III, 527.	Phenol derivative of unknown structure
(7) 27, 168, 870; I, 266; II, 214.	Ambiguity: do not use Subject Index
(8) 9, 129; II, 109; III, 536.	
(9) 2, III, 1023	⎱ Cannot be located by indexes;
(10) 2, III, 449.	⎰ use the system. No entries in *EIV*
(11) 2, III, 575.	
(12) 12, 1045; I, 458; III, 2259.	
(13) 31, 424; 17, III/IV, 3786	Natural product of unknown structure 31/12/09
(14) 2, 601; I, 259; II, 540; III, 1643; IV, 1908.	Indexed under Bersteinsäure
(15) 17, III/IV, 2905.	
(16) 1, IV, 123.	

Exercise 9

(1) (a) and (b) See Heilbron and Bunbury's *Dictionary of Organic Compounds*.
    (c) See *Beilstein*, Band 24.
    (d) See *Organic Syntheses* or *Advances in Protein Chemistry*, 1957.
(2) See *Beilstein*, **3**, 125; I, 59; II, 99; III, 236; IV, 248
(3) See Weissberger, Vol. 7, or Faraday's *Encyclopedia*.
(4) Consult *Organic Reactions* (Volume 9). For melting points consult Heilbron and Bunbury's *Dictionary of Organic Compounds*.
(5) It is probably easier to use the *Ring Index* than the *Colour Index*.
(6) See Chapter 18: 'The detailed survey'.

Exercise 10

(1 and 2) Procedure is self-evident.
(3) Birmingham, Pergamon, 1960 (from *British National Bibliography*).
(4) See Bertolacini and Barney, *Analytical Chemistry*, 1957, **29**, 281–3. For a detailed discussion of this question, see Reference to Ridland (1960), Chapter 18.

Exercise 11

(1) (a) Go first to the *British Patent Specifications* themselves to find

what range of numbers was being given to patents published about 1952 (approximate range 660000–680000). Then consult the volume *Name Indexes to Complete Specifications* which covers this range. (The volumes cover 20000 specifications each, so for this question the volume covering 660001 to 680000 should be consulted). From the entries under 'Distillers Co. Ltd' determine which is likely to be patent referred to in question (BP 677091). Consult specification itself for confirmation and for scope of claim. Answer to second part of question – No.

(b) Procedure similar to (a). Patent No. 674896. Answer to second part of question – No.

(c) Procedure similar to (a) except that volumes covering ranges 760001 to 780000 and 780001 to 800000 would be consulted. Patent No. 770264. Answer to second part of question – Yes.

(d) Procedure similar to (c). Patent No. 782066. Answer to second part of question – Yes.

(2) Locate US specification – make note of application date, inventors' names, name of company to whom patent rights assigned (all these given on front page of specification) and essential features of main claim. Then consult *British Patent Specifications* to find what numbers were being allocated to British Specifications at application date of US patent. Then refer to British *Name Indexes to Complete Specifications*, from that number onwards, searching under name of assignee company (inventors' names will usually be helpful under the company entry) for any hopeful-sounding titles. Then consult British specification for confirmation, comparing features of main claims, bearing in mind that complete correspondence is unlikely.

Answers: (a) BP 782322, (b) BP 791995, (c) BP 783111, (d) BP 785387, (e) BP 810525, (f) BP 783027, (g) BP 799421.

(3) Refer to *British Patent Specifications* to determine range of numbers published in the period specified in the question. Then consult *Classification Key Reference Index* to determine the class reference of the subject being sought. Having found the class from the main part of the *Index*, consult the back of it to find in which group that class falls. Then refer to the volumes of *Abridgements of Specifications* covering the group found and the range of numbers first determined. In the front of each volume is a subject index of all classes found in that volume, listing the numbers of all specifications dealing with each subject. Locate appropriate heading and then consult the abridgement of each number listed to see if it is required to answer question.

Answers: (a) 648 995; 654 692; 658 188; 658 189; 670 906; 673 273; 680 508
(b) 650 438; (651 203); 654 764; 655 237; 655 439; 658 954.
(c) 665 698; 666 785; 679 538; 683 898.
(d) (685 522); 687 843; (689 607); 690 008; 704 706; 711 657; (720 415).
(e) 648 752; 649 286; 649 945; 654 786; (655 690); (655 715); 657 550.
(f) 666 554; 666 589; 666 720; 669 074; 670 444; (670 445); 672 512; (673 576); (674 754); (675 193); 676 109; 676 770; 676 771; (678 192); 679 411; 679 826.

Numbers in parentheses refer to patents of marginal relevance.

(4) Procedure is self-evident.

Exercise 12

Consultation of the references listed below will provide answers to the queries. Where publication details are not given, the reference has been described previously.

(1) Sax, *Dangerous Properties of Industrial Materials*.
(2) *Handbook of Chemistry and Physics*, 43rd edn, p. 1450.
(3) *J. Chem. Phys.*, 1957, **26**, 1636 (via *Chem. Abstr.*).
(4) *International Critical Tables*, Vol. 5, p. 100.
(5) *Merck Index*.
(6) Kirk–Othmer's *Encyclopedia of Chemical Technology*.
(7) *ICT*.
(8) Keesom, *Helium* (Elsevier, 1959).
(9) Von Loesecke, *Drying and Dehydration of Foods* (2nd edn, Reinhold, 1955). N.B.: This is a case where it would probably be quicker to make an experimental determination (e.g. Dean and Stark method) of the percentage moisture in raisins than to look it up.
(10) *Handbook of Chemistry and Physics*, 43rd edn, p. 2130. (Little error is introduced by assuming specific gravity is the same as the equivalent ethanol/water mixture.)
(11) (a) Bailey, *Industrial Oil and Fat Products* (2nd edn, Interscience, 1951);
(b) Kirk–Othmer's *Encyclopedia*, Vol. 9.
(12) Thorpe's *Dictionary of Applied Chemistry* (under 'Wines');
Von Loesecke, *Drying and Dehydration of Foods*;
Spiers, *Technical Data on Fuel* (6th edn, World Power Conference, 1961).
(15) Brown and Zerban, *Sugar Analysis* (Wiley, 1941).
(16) *Chem. Abstr.* Subject Indexes.
(17) *Chem. Abstr.*, 37, 5337[8].
(18) Kirk–Othmer's *Encyclopedia*, Vol. 9.

(19) Timmermans, *Physicochemical Constants of Pure Organic Compounds* (Elsevier, 1960).
(20) Kirk–Othmer's *Encyclopedia*.
(21) *Chemist and Druggist Yearbook*.
(22) Lange's *Handbook of Chemistry*.
(23) Thorpe's *Dictionary of Applied Chemistry*.
(24) *BP Codex*, 1973, p. 510 (or *Merck Index*).

Exercise 13

See Chapter 5, Table 5.1.

Exercise 14

Solution is self-evident.

Exercise 15

(1) 1VOV1.
(2) 2OVM1VO2.
(3) $CH_3CH_2CH_2C{\equiv}CCH_2CH_3$.
(4) $NH_2CH_2CH_2COOH$.
(5) $CH_3SCH_2CH_2OOCCH_2NH_2$.
(6) ZR CM1.
(7) Q2R D1U1R BQ.

(8)

$-CH_2CH_2CH_3$

(9)

$-CH_2CO-$

(10)

$-NHNH_2$

Cl

(11)

$-NH-$

HO NH_2

(12)

Index

Only the more important reference books and journals have been indexed by name and then only where there is descriptive matter pertaining to them. No entries have been made to pages where they are merely cited in lists or as examples. Because of lack of space in this edition for the Glossary of Acronyms, etc., acronyms and initials have frequently been used as index entries instead of the full name of the organization, etc. The index will therefore lead to the page where the acronym is defined. It follows therefore that if the reader cannot find the organization, title, etc., under its full form, he should look under its initials or acronym.